Apache Spark for Machine Learning

Build and deploy high-performance big data AI solutions for large-scale clusters

Deepak Gowda

Apache Spark for Machine Learning

Group Product Manager: Niranjan Naikwadi

Publishing Product Manager: Nitin Nainani

Book Project Manager: Aparna Nair

Senior Editor: Rohit Singh

Technical Editor: Rahul Limbachiya

Copy Editor: Safis Editing

Proofreader: Rohit Singh

Indexer: Hemangini Bari

Production Designer: Alishon Mendonca

DevRel Marketing Executive: Vinishka Kalra

First published: November 2024

Production reference: 1260924

Published by Packt Publishing Ltd.

Grosvenor House

11 St Paul's Square

Birmingham

B3 1RB, UK.

ISBN 978-1-80461-816-5

www.packtpub.com

Contributors

About the author

Deepak Gowda is a data scientist and AI/ML expert with over a decade of experience in leading innovative solutions across various industries, including supply chain, cybersecurity, and data center infrastructure. He holds over 30 granted patents, contributing to advancements in automation, predictive analytics, and AI-driven optimization. His work spans data engineering, machine learning, and distributed systems, focusing on building scalable and impactful products. A passionate inventor, mentor, author, and FAA-certified pilot, Deepak is also dedicated to content creation, sharing his expertise through writing, speaking, and mentoring. He continues to push the boundaries of technology, driving innovation across sectors.

About the reviewers

Karthik Hubli is a backend and AI engineer with over a decade's experience at leading global companies such as Infosys, Thomson Reuters, StateStreet, and Dell Technologies, with expertise in distributed systems, ML, and AI. He has made significant contributions to the fields of time series forecasting, log parsing, and NLP, which has been recognized with several patents, underscoring his innovative impact on the industry. He is a leader in the development of SaaS platforms. Karthik is also an accomplished author, having published an e-book on LLMs and another book focused on cutting-edge developments in ML and AI on Amazon. His work as a book reviewer with Packt Publishing adds valuable insights that shape the future of AI and data science.

Siva Rama Krishna Kottapalli is a seasoned data scientist and software engineer with over 9 years of experience, having developed a passion for harnessing the power of deep learning technologies and data to drive business growth. His expertise spans transformer models, LLM-based natural language processing, and deep learning techniques. Siva's notable work includes the development of a cutting-edge **Retrieval-Augmented Generation (RAG)** system, integrating state-of-the-art language models with a vector database for efficient knowledge retrieval and contextual response generation. He has developed robust security solutions and architected scalable data platforms to enable secure data exchange and real-time insights.

Table of Contents

5

Building a Classification System   127

Part 3: Unsupervised Learning

6

Building a Clustering System   169

7

Building a Recommendation System | 205

8

Mining Frequent Patterns | 231

Part 4: Model Deployment

9

Deploying a Model   249

Preface

Welcome to this journey into the world of machine learning with Apache Spark, a powerful combination that is revolutionizing the way data is processed and analyzed in today's fast-paced digital environment. This book is designed to provide you with a comprehensive understanding of machine learning concepts, coupled with practical insights into how these concepts can be implemented using Apache Spark, a leading platform for big data processing.

As the need for processing large volumes of data continues to grow across various industries, the ability to efficiently apply machine learning techniques at scale has become increasingly critical. This book is crafted with the intention of bridging the gap between theory and practice, offering not only a solid grounding in the fundamentals of machine learning but also practical examples of how to harness the capabilities of Apache Spark, allowing to build and deploy sophisticated models.

Who this book is for

This book is for data scientists, machine learning engineers, and software developers who are eager to expand their knowledge and skills in machine learning, particularly in the context of big data. Whether you are a seasoned professional or a newcomer to the field, this book provides valuable insights into both the theoretical underpinnings of machine learning and the practical application of these principles, using Apache Spark.

The content is tailored to those who are looking to enhance their ability to work with large datasets and deploy machine learning models in real-world scenarios. If you are interested in understanding how to integrate machine learning into your big data workflows, this book will serve as a comprehensive guide.

What this book covers

Chapter 1, *An Overview of Machine Learning Concepts*, introduces you to the core principles of machine learning, including supervised, unsupervised, and reinforcement learning, and their relevance in the context of Apache Spark. You will also learn how to set up Apache Spark on your local machine to begin exploring its capabilities.

Chapter 2, *Data Processing with Spark*, delves into the critical processes of data preprocessing, transformation, and cleaning, which are essential for preparing your data for effective machine learning. The chapter covers key techniques and best practices for handling large datasets in Spark.

Chapter 3, Feature Extraction and Transformation, focuses on the techniques used to convert raw data into a format suitable to train machine learning models. You will explore various feature extraction and transformation methods, learning how to implement these in Spark.

Chapter 4, Building a Regression System, dives into the world of regression analysis, a fundamental technique in supervised learning. This chapter guides you through building and evaluating regression models using Spark's machine learning library.

Chapter 5, Building a Classification System, covers classification, explaining key algorithms such as decision trees, logistic regression, and random forests. It includes practical examples, performance evaluation, and improvement strategies.

Chapter 6, Building a Clustering System, explores unsupervised learning through clustering algorithms, such as *K*-means and hierarchical clustering. This chapter provides practical examples of how to implement these techniques in Spark.

Chapter 7, Building a Recommendation System, focuses on the practical application of machine learning in developing recommendation systems, an essential tool in many industries today. You will learn how to use collaborative filtering and other techniques to build a recommendation system in Spark.

Chapter 8, Mining Frequent Patterns, discusses the techniques used in frequent pattern mining – a crucial task in data mining – and shows how to apply these methods using Apache Spark to uncover hidden patterns in large datasets.

Chapter 9, Deploying a Model, covers the deployment phase of machine learning models, emphasizing the importance of operationalizing data-driven insights. You will learn about various deployment strategies, model monitoring, and how to ensure that your models perform effectively in a production environment.

To get the most out of this book

To fully benefit from this book, it is recommended that you have a basic understanding of machine learning concepts and some familiarity with Apache Spark. Throughout the book, you will find practical examples, code snippets, and step-by-step instructions that are designed to be easily followed and applied to your own projects.

To follow the instructions in this book, you will need the following:

Software/hardware covered in the book	Operating system requirements
Python 3.x	Windows, macOS, or Linux
Apache Spark 3.x.x	
ECMAScript 11	

If you are using the digital version of this book, we advise you to type the code yourself or access the code from the book's GitHub repository (a link is available in the next section). Doing so will help you avoid any potential errors related to the copying and pasting of code.

Downloading the example code files

The following is the list of steps to download and run the code:

1. **Clone or download the repository**: You can either clone the repository to your local machine using Git or download the ZIP file containing all the code files. If you are familiar with Git, cloning the repository will allow you to pull updates easily. To clone the repository, use the following command:

    ```Bash
    git clone https://github.com/PacktPublishing/Apache-Spark-for-
    Machine-Learning.git
    ```

2. **Navigate to the desired chapter**: Each chapter has its own folder within the repository. Navigate to the appropriate folder to find the code files related to the chapter you are reading.

3. **Run the code examples**: The code examples are written in Python, and most of them require Apache Spark to be installed on your machine. Follow the setup instructions provided in the book to ensure that your environment is correctly configured. You can run the code examples directly in your Python or PySpark environment.

4. **Modify and experiment**: Feel free to modify the code to experiment with different parameters or datasets. This will help you better understand the concepts and see how changes impact the results.

You can download the example code files for this book from GitHub at `https://github.com/PacktPublishing/Apache-Spark-for-Machine-Learning`.

If there's an update to the code, it will be updated in the GitHub repository.

We also have other code bundles from our rich catalog of books and videos available at `https://github.com/PacktPublishing/`. Check them out!

Conventions used

There are a number of text conventions used throughout this book.

`Code in text`: Indicates code words in text, database table names, folder names, filenames, file extensions, pathnames, dummy URLs, user input, and Twitter handles. Here is an example: "Ensure that the `JAVA_HOME` environment variable is properly set."

A block of code is set as follows:

```
from pyspark.sql import SparkSession
spark = SparkSession.builder \
    .appName("HDFS Read Example") \
    .getOrCreate()
```

Any command-line input or output is written as follows:

```
tar -xvf spark-3.x.x-bin-hadoop3.x.tgz
```

Bold: Indicates a new term, an important word, or words that you see on screen. For instance, words in menus or dialog boxes appear in **bold**. Here is an example: "For local development, you can choose the **Pre-built for Apache Hadoop** option."

> **Tips or important notes**
> Appear like this.

Get in touch

Feedback from our readers is always welcome.

General feedback: If you have questions about any aspect of this book, email us at customercare@packtpub.com and mention the book title in the subject of your message.

Errata: Although we have taken every care to ensure the accuracy of our content, mistakes do happen. If you have found a mistake in this book, we would be grateful if you would report this to us. Please visit www.packtpub.com/support/errata and fill in the form.

Piracy: If you come across any illegal copies of our works in any form on the internet, we would be grateful if you would provide us with the location address or website name. Please contact us at copyright@packt.com with a link to the material.

If you are interested in becoming an author: If there is a topic that you have expertise in and you are interested in either writing or contributing to a book, please visit authors.packtpub.com.

Share Your Thoughts

Once you've read *Apache Spark for Machine Learning*, we'd love to hear your thoughts! Scan the QR code below to go straight to the Amazon review page for this book and share your feedback.

https://packt.link/r/1-804-61816-0

Your review is important to us and the tech community and will help us make sure we're delivering excellent quality content.

Download a free PDF copy of this book

Thanks for purchasing this book!

Do you like to read on the go but are unable to carry your print books everywhere?

Is your eBook purchase not compatible with the device of your choice?

Don't worry, now with every Packt book you get a DRM-free PDF version of that book at no cost.

Read anywhere, any place, on any device. Search, copy, and paste code from your favorite technical books directly into your application.

The perks don't stop there, you can get exclusive access to discounts, newsletters, and great free content in your inbox daily

Follow these simple steps to get the benefits:

1. Scan the QR code or visit the link below

https://packt.link/free-ebook/9781804618165

2. Submit your proof of purchase

3. That's it! We'll send your free PDF and other benefits to your email directly

Part 1:
Introduction and Fundamentals

In this part, you will embark on a journey through the foundational concepts and principles that underpin machine learning and its integration with Apache Spark. This part is designed to equip you with the essential knowledge required to navigate the more advanced topics covered later in the book.

The chapters in this part will introduce you to the core principles of machine learning, the architecture and capabilities of Apache Spark, and the techniques for feature extraction and transformation. By the end of this part, you will have a solid understanding of both the theoretical and practical aspects of these fundamental concepts.

This part contains the following chapters:

- *Chapter 1, An Overview of Machine Learning Concepts*
- *Chapter 2, Data Processing with Spark*
- *Chapter 3, Feature Extraction and Transformation*

1

An Overview of Machine Learning Concepts

This chapter provides a comprehensive introduction to the integration of machine learning within the Apache Spark ecosystem. It begins by elucidating fundamental machine learning principles, such as supervised, unsupervised, and reinforcement learning, and their relevance to Spark's distributed computing paradigm. You will gain insights into its rich set of algorithms for classification, regression, clustering, and recommendation tasks. Furthermore, the chapter elucidates why Spark is used for machine learning, examining its use cases and benefits. It will also help you to set up Apache Spark on a local machine.

We will cover the following topics in this chapter:

- Understanding machine learning
- An introduction to Apache Spark
- Why Apache Spark for machine learning?
- Setting up Apache Spark

By the end of this chapter, you will know the basics of machine learning, Apache Spark, and how to set it up.

Technical requirements

To run Apache Spark on a local machine, you typically need the following technical requirements:

- **Operating system**: Apache Spark is compatible with Linux, macOS, and Windows.
- **Java Development Kit (JDK)**: Apache Spark is implemented in Java, so you need to have JDK installed. Ensure that the JAVA_HOME environment variable is properly set.
- **Python**: If you plan to use PySpark (the Python API for Apache Spark), you'll need to have Python installed. Python 3.x is recommended.

You can find the code files for this chapter on GitHub at `https://github.com/PacktPublishing/Apache-Spark-for-Machine-Learning/tree/main/Chapter01`.

Understanding machine learning

We will begin with a gentle introduction to machine learning. **Machine learning** (**ML**) is a branch of **artificial intelligence** (**AI**). It focuses on developing algorithms and techniques that enable computers to learn from data and improve their performance on specific tasks over time, all without being explicitly programmed. At its core, machine learning is about extracting patterns and insights from data to make predictions or decisions.

There are several key paradigms within machine learning:

- **Supervised learning**: This involves training a model on labeled data, where the algorithm learns to map input data to corresponding output labels. It's used for tasks such as classification and regression.

- **Unsupervised learning**: This involves training a model on unlabeled data, where the algorithm learns to find hidden patterns or structures within the data. It's used for tasks such as clustering and dimensionality reduction.

- **Reinforcement learning**: This involves training a model to make decisions sequentially through interaction with an environment, by receiving feedback in the form of rewards or penalties.

Machine learning algorithms can be further categorized based on their functionality, such as decision trees, neural networks, and support vector machines.

The success of machine learning relies heavily on data quality, quantity, and relevance. Additionally, factors such as feature engineering, model selection, hyperparameter tuning, and evaluation metrics play crucial roles in developing effective machine learning systems.

Machine learning can be found applied across diverse domains, including healthcare, finance, marketing, autonomous vehicles, and recommendation systems. Its ability to analyze large volumes of data, identify complex patterns, and make data-driven predictions empowers organizations to gain valuable insights, optimize processes, and make informed decisions, driving innovation and transformation in today's data-driven world.

Imagine you have a baby and want to teach it to recognize different objects. This process is like supervised machine learning.

Let's understand the flow of a machine learning solution:

- **Training data**: In machine learning, this is like giving a baby a set of toys and telling it what each toy is. For example, you show the baby a ball and say, "*This is a ball,*" and then you show it a bone and say, "*This is a bone.*"

In machine learning, we provide a computer algorithm with a dataset consisting of examples (input data such as an image) and their corresponding labels (the output or desired outcome, such as a dog or cat).

- **Learning algorithm**: The learning algorithm is the baby's brain. It learns by observing and understanding the features of each object.

  In machine learning, the algorithm processes the training data and learns patterns and features that help it make predictions or classifications.

- **Testing and evaluation**: Now, you present new objects to the baby that it hasn't seen before, such as a football or a balloon, and see whether it can correctly identify them based on what it learned.

  In machine learning, you test the trained algorithm on a separate dataset (testing data) to evaluate its performance and see how well it generalizes to new, unseen examples.

- **Adjustments and iterations**: If the baby makes mistakes, you might correct it by saying, *"No, that's not a ball; it's a football."* The baby learns from these corrections.

  In machine learning, if the algorithm makes errors, you adjust its parameters or even the features it considers to improve its accuracy. This process may involve multiple iterations.

- **Model deployment**: Once the baby consistently identifies objects correctly, it can be said to have been "deployed" successfully as a reliable object recognizer.

  In machine learning, when the algorithm performs well on the testing data and meets the desired accuracy, it can be deployed to make predictions on new real-world data.

  This example helps to simplify the complex process of machine learning by drawing parallels to a familiar scenario involving learning and recognition. Remember that this is a simplified representation, and machine learning involves various algorithms, models, and techniques that can be much more sophisticated.

Three key ingredients are required to build machine learning models. Let's look at those three important components:

- Data
- Algorithms
- Hardware

Data is a fundamental and critical component in machine learning. It is the raw material from which a machine learning model learns patterns, makes predictions, and gains insights.

Understanding the characteristics and quality of data is crucial for the success of a machine learning project. High-quality, relevant, and representative data contributes significantly to a model's generalization of new unseen examples. More details are covered in *Chapter 2, Data Processing with Spark*. Here are some key aspects of data in machine learning:

- **Quality data**: Quality data is clean, structured, representative, balanced, labeled, and sufficient. For example, a quality dataset for image classification would have clear and consistent images of different objects, with a diverse and balanced distribution of classes and accurate labels for each image. Quality data helps ML models to learn effectively and perform well on new data.

- **Poor quality data**: Poor quality data is noisy, inconsistent, biased, imbalanced, missing, or duplicated. For example, a poor-quality dataset for sentiment analysis would have corrupted text data, incomplete or containing spelling errors, a skewed or unrepresentative sample of opinions, and missing or incorrect labels for each text.

Machine learning algorithms are mathematical models or computational procedures that enable computers to learn from data and make predictions or decisions, without being explicitly programmed. These algorithms form the core of machine learning systems and are designed to discover patterns, relationships, and insights within data. Understanding different algorithms' characteristics, strengths, and limitations is crucial in selecting the most appropriate one for a specific machine learning task. The choice of algorithm depends on factors such as the nature of the data, the problem at hand, and the available computational resources. Refer to *Table 1.1* later in this section for more details.

Hardware too plays a crucial role in machine learning, influencing the speed, efficiency, and scale at which models can be trained and deployed. The choice of hardware depends on the complexity of the machine learning task, the dataset size, and the algorithms' computational requirements. Choosing the right hardware depends on factors such as the dataset's size, the model's complexity, and the machine learning task's specific requirements. The field continues to evolve with ongoing advancements in hardware architectures and technologies.

Types of machine learning

Machine learning can be broadly categorized into three main types, based on the learning approach and the nature of the training data:

- Supervised learning
- Unsupervised learning
- Reinforcement learning

These types can be further sub-classified as follows:

- Semi-supervised learning
- Transfer learning

- Ensemble learning
- Deep learning

We will now discuss each of these main types in detail.

Supervised learning

In supervised learning, the algorithm is trained on a labeled dataset, and input is paired with the corresponding desired label. The goal of supervised learning is to develop a predictive model that can accurately map input data to desired output labels, making accurate predictions or classifications on new unseen data.

Use cases of supervised learning include the following:

- **Classification**: Predicting whether an email is spam
- **Regression**: Predicting the price of a house based on its features

The following diagram shows an example of supervised model training and its output:

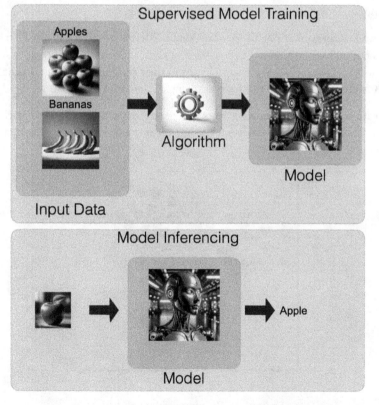

Figure 1.1 – An example of supervised learning

For example, if you want to classify fruits such as apples and bananas, you need a dataset of images of apples and bananas, where each image has a label indicating what fruit it is. The algorithm then learns to recognize the features that distinguish apples from bananas and can predict the label for new images it has not seen before. One way to perform supervised learning for apples and bananas is to use a **convolutional neural network (CNN)**.

Unsupervised learning

Unsupervised learning involves training an algorithm on an unlabeled dataset, where the algorithm must discover patterns and relationships in data without explicit guidance.

The goal is often to identify hidden structures, group similar data points, or reduce the dimensionality of the data.

Use cases of unsupervised learning include the following:

- **Clustering**: Grouping customers based on their purchasing behavior
- **Dimensionality reduction**: Reducing the number of features in a dataset

K-means is a popular algorithm used for unsupervised learning.

Imagine you have a dataset containing information about different types of customers in a retail store, but the data does not include any labels or categories for these customers. You want to segment these customers into distinct groups, based on similarities in their purchasing behaviors, demographics, or other relevant features, but you do not have predefined categories for them.

A K-means algorithm outputs a set of K clusters that group customers with similar characteristics together. The following diagram shows several clusters resulting from unsupervised training:

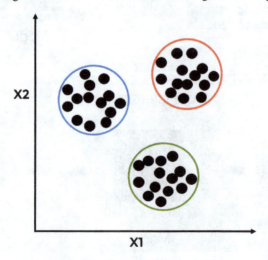

Figure 1.2 – Unsupervised training

In this diagram, there are several data points each indicated by a small circle. Data points with similar characteristics are grouped together, using a different color for each cluster

Reinforcement learning

Reinforcement learning is a type of machine learning where an agent learns to make decisions by performing actions in an environment to achieve a goal. The agent learns from the consequences of its actions, rather than from explicit instruction, through a system of rewards and punishments. The agent aims to learn a policy that maximizes the cumulative reward over time.

Use cases of reinforcement learning include the following:

- **Game playing**: Training a computer program to play chess or Go
- **Robotics**: Teaching a robot to perform tasks in the physical world

A classic example of reinforcement learning is training an agent to play a game, such as chess. In this example, the environment is the chessboard, and the agent is the player making decisions on which moves to make. The goal of the agent is to win the game. The agent starts with little to no knowledge about the game. It does not know the best moves but learns through trial and error. After each move, the agent receives a reward (good moves) or a penalty (bad moves). After numerous games and learning from the outcomes of various actions, the agent becomes proficient at playing chess.

Figure 1.3 – Reinforcement learning

These main types of machine learning can further be broken down into subtypes and specialized approaches. Some additional categories include the following:

- **Semi-supervised learning**: Combines elements of both supervised and unsupervised learning. The model is trained on a dataset containing labeled and unlabeled examples.

- **Transfer learning**: Involves training a model on one task and then applying the learned knowledge to a different but related task.

- **Ensemble learning**: Involves combining multiple models to improve overall performance and robustness.

- **Deep learning**: Utilizes neural networks with multiple layers (deep neural networks) to automatically learn hierarchical representations of data. It is a subset of machine learning and is particularly effective for tasks such as image and speech recognition.

Understanding these types of machine learning is crucial for selecting the appropriate approach for a given task or problem. Different types may be more suitable, depending on the nature of the data, the available resources, and the specific goals of the machine learning application.

Table 1.1 discusses various algorithms used for different ML applications and some popular use cases:

Type of machine learning	Algorithm examples	Use cases
Supervised learning	<ul><li>Linear regression</li><li>Decision trees</li><li>**Support Vector Machines (SVMs)**</li><li>Neural networks</li></ul>	<ul><li>Image classification</li><li>Stock price prediction</li><li>Spam email detection</li></ul>
Unsupervised learning	<ul><li>K-means clustering</li><li>Hierarchical clustering</li><li>**Principal Component Analysis (PCA)**</li></ul>	<ul><li>Customer segmentation</li><li>Anomaly detection</li><li>Dimensionality reduction</li></ul>
Reinforcement learning	<ul><li>Q-learning</li><li>**Deep Q Networks (DQNs)**</li><li>Policy gradient methods</li></ul>	<ul><li>Game playing (for example, AlphaGo)</li><li>Robotics</li><li>Autonomous systems</li></ul>
Semi-supervised learning	<ul><li>Self-training</li><li>Co-training</li><li>Multi-view learning</li></ul>	<ul><li>Text and speech processing</li><li>Image recognition</li></ul>

Transfer learning	• Fine-tuning pre-trained models • Feature extraction • Domain adaptation	• Image recognition (for example, using pre-trained CNNs) • **Natural Language Processing (NLP)**
Ensemble learning	• Random forests • **Gradient Boosting Machines (GBMs)** • AdaBoost	• Improved classification accuracy • Robust predictions
Deep learning	• CNNs • **Recurrent Neural Networks (RNNs)** • Transformer models (for example, GPT)	• Image and speech recognition • NLP • Deep reinforcement learning

Table 1.1 – Algorithms and their use cases

So far, we have focused on learning the fundamentals of machine learning and some of its popular algorithms. We will now shift our focus to learning about Apache Spark.

An introduction to Apache Spark

Apache Spark is a powerful, open source, unified analytics engine, designed for large-scale data processing and machine learning tasks. It provides high-level APIs in Java, Scala, Python, and R and has an optimized engine that supports general computation graphs for data analysis, offering speed and ease of use for developers. Spark's core functionality, coupled with its libraries for SQL, streaming, machine learning, and graph processing, makes it a versatile tool for a wide range of data processing and analytics tasks, from batch processing to real-time analytics and machine learning.

The background and motivation of Apache Spark

In the era of big data, the need for scalable, fast, and flexible data processing frameworks became increasingly apparent. Traditional solutions, such as Apache Hadoop MapReduce (https://en.wikipedia.org/wiki/MapReduce), paved the way for distributed data processing but fell short in speed and ease of use. In response to these challenges, researchers conceived Apache Spark as a revolutionary open source project at UC Berkeley's AMPLab in 2009.

This section delves into the background and motivations behind the creation of Apache Spark, highlighting its evolution as a powerful data processing framework.

Challenges with MapReduce

Apache Hadoop MapReduce, while groundbreaking, had limitations that hindered its widespread adoption. The disk-based nature of intermediate data storage and the necessity to write to disk after each map and reduce operation introduced latency, impacting the overall processing speed. Additionally, the complex and verbose nature of MapReduce programs, including the other following challenges, made it less developer-friendly:

- **Complexity and verbosity**: Implementing MapReduce programs can be complex and verbose. Developers must write code for both the map and reduce phases, which can lead to a substantial amount of boilerplate code.

- **Programming paradigm**: MapReduce follows a functional programming paradigm, which can be unfamiliar for developers accustomed to more traditional imperative programming languages. This shift in thinking may pose a learning curve for some developers.

- **Data movement overhead**: The shuffling and sorting phases in MapReduce involve significant data movement across a network, which can lead to overhead. Efficiently managing and minimizing data movement is crucial for optimizing performance.

- **Limited support for iterative algorithms**: MapReduce is not well-suited for iterative algorithms, which are common in machine learning and graph processing tasks. Running multiple MapReduce jobs for iterative algorithms introduces additional complexity and performance overhead.

- **Debugging and testing**: Debugging MapReduce jobs can be challenging. Traditional debugging tools may not be as effective in a distributed environment. Developers often resort to log analysis and custom debugging techniques.

- **Limited support for real-time processing**: MapReduce is designed for batch processing and may not be the best choice for real-time or low-latency processing requirements. Other frameworks, such as Apache Spark, have emerged to address these use cases.

Apache Spark was developed to address limitations in the MapReduce processing model, which was the primary data processing framework within the Apache Hadoop ecosystem.

Here is a brief timeline of how Apache Spark came into existence:

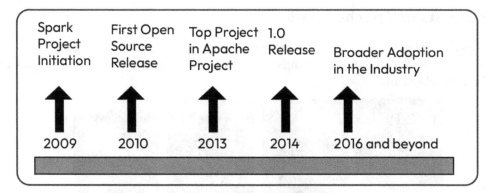

Figure 1.4 – A Spark development timeline

Today, Apache Spark is widely used for large-scale data processing, machine learning, graph analytics, and so on. Its versatility, speed, and support for in-memory processing have contributed to its popularity, and it has become a foundational component in modern big data architectures. Apache Spark is an open source, distributed computing system that provides a fast and general-purpose cluster computing framework for big data processing.

Let's discuss the key features of Apache Spark that make it exciting to use for ML:

- **In-memory processing**: One of the key differentiators of Apache Spark is its in-memory processing capabilities. By storing intermediate data in memory rather than persisting it to disk, Spark dramatically accelerates iterative algorithms and interactive data analysis. This approach enhances performance and facilitates complex computations on large datasets.

- **Unified processing engine**: Apache Spark provides a unified processing engine for batch and stream processing, machine learning, graph processing, and SQL queries. This versatility eliminates the need for separate tools for different tasks, streamlining the development process and reducing the user learning curve.

- **Fault tolerance**: Spark's fault tolerance mechanisms are crucial for maintaining data integrity in distributed computing environments. Spark can reconstruct lost data through lineage information if there are node failures. This resilience ensures the reliability of Spark applications even in large-scale and dynamic clusters.

- **Ease of use**: Designed with user-friendliness in mind, Apache Spark offers high-level APIs in Java, Scala, Python, and R. This design choice broadens its accessibility, enabling data engineers and data scientists to leverage its capabilities. Spark's concise syntax and interactive shell provide a more intuitive user experience.

Next, we will discuss the architecture of Apache Spark and how various components interact with each other.

Components of Apache Spark

Apache Spark consists of several components that work together to provide a comprehensive and unified engine for big data processing and analytics. The following diagram shows various components that exist in Apache Spark:

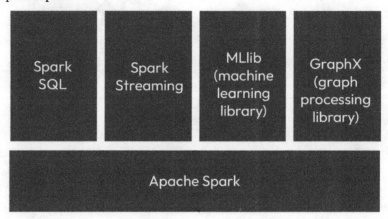

Figure 1.5 – Spark components

Let us learn what each component does:

- **Spark Core**: At the core of Apache Spark lies Spark Core, providing essential functionality for the entire Spark ecosystem. It includes task scheduling, memory management, and fault recovery, forming the foundation for other Spark libraries.

- **Spark SQL**: Spark SQL facilitates querying structured data using SQL commands. This component seamlessly integrates SQL queries with Spark programs, catering to data analysts and SQL-savvy users for whom SQL is a familiar and efficient tool.

- **Spark Streaming**: Addressing the need for real-time analytics, Spark Streaming processes data streams in near-real-time using micro-batch processing. It enables the application of Spark's powerful batch-processing capabilities to streaming data sources.

- **MLlib (Machine Learning Library)**: MLlib, Spark's machine learning library, offers scalable implementations of various machine learning algorithms. With MLlib, data scientists can build and deploy machine learning models on large datasets, leveraging Spark's in-memory processing for faster training.

- **GraphX**: GraphX is Spark's graph processing library, providing a resilient distributed graph system. It enables you to create and manipulate graphs, making it well-suited for complex relationships and network analysis applications.

Use cases and applications of Apache Spark

Let's look at a list of use cases and real-world examples of Spark:

- **Large-scale data processing**: One of the primary use cases for Apache Spark is large-scale data processing. Organizations deploy Spark for tasks such as log analysis, data cleaning, and **Extract, Transform, and Load** (ETL) operations, benefiting from its efficient and scalable processing capabilities. For example, Apache Spark can be useful in processing logs, as they are highly complex and involve terabytes of data.

- **Real-time analytics**: Spark Streaming's ability to process real-time data streams positions Apache Spark as a key player in the realm of real-time analytics. It caters to businesses seeking timely insights from streaming data sources, such as social media, sensors, and logs. A use case for real-time analytics is the processing of incoming telemetry data to detect anomalies.

- **Machine learning at scale**: MLlib empowers data scientists to build and deploy machine learning models at scale. Spark's in-memory processing capabilities significantly accelerate the training of complex models, making it an ideal choice for organizations with large datasets. For example, MLlib can be used to train ML models on datasets containing billions of rows, such as sensor data to monitor the health of a data center.

In the next section, we will examine the advantages of Apache Spark and its collection of machine learning algorithms.

Why Apache Spark for machine learning?

Apache Spark offers several advantages for machine learning applications, making it a popular choice for scalable and distributed ML tasks. Here are some key advantages of using Apache Spark for machine learning:

- **In-memory processing**: Spark's ability to store intermediate data in memory accelerates iterative algorithms commonly used in machine learning, significantly reducing computation time.

- **Distributed computing**: Spark's distributed computing capabilities allow for the parallel processing of large datasets across a cluster of machines, enabling scalability for ML tasks.

- **Resilient Distributed Datasets (RDDs)**: Spark's fundamental data structure, RDDs, provides fault-tolerant parallel processing. In the context of machine learning, this means that if a node fails, the computation can continue on other nodes without losing progress.

- **Unified platform**: Spark provides a unified platform for data processing and machine learning, eliminating the need for separate tools. This simplifies the overall workflow and enhances the ease of integration.

- **Ease of use**: Spark offers high-level APIs in Java, Scala, Python, and R, making it accessible to many developers and data scientists. This ease of use facilitates the development and deployment of machine learning models.

- **MLlib**: MLlib, Spark's machine learning library, provides a rich set of algorithms and tools for machine learning tasks. It includes scalable classification, regression, clustering, collaborative filtering, and more implementations.

- **Data processing capabilities**: Spark's capabilities for data preprocessing, cleaning, and transformation are seamlessly integrated with its machine learning libraries. This streamlines the end-to-end process of building and deploying machine learning models.

- **Streaming integration**: Spark Streaming allows the integration of real-time data streams into machine learning pipelines. This is crucial for applications requiring real-time predictions or continuous model updates.

- **Graph processing (GraphX)**: For machine learning tasks involving graph-structured data, Spark's GraphX library simplifies the development of graph algorithms and analytics, allowing for the integration of graph processing into ML workflows.

- **Community support and ecosystem**: Spark benefits from a vibrant open source community and a growing ecosystem of libraries and tools. This provides additional resources and support for developers and data scientists working on machine learning projects.

- **Compatibility with the Hadoop ecosystem**: Spark can run on Hadoop clusters, leveraging existing Hadoop infrastructure. This compatibility makes it easier for organizations with Hadoop deployments to adopt Spark for machine learning without major architectural changes.

- **Performance optimizations**: Spark incorporates various performance optimizations, including caching, pipelining, and query optimization, contributing to improved efficiency in machine learning computations.

Algorithms in Apache Spark

Apache Spark's machine learning library, MLlib, provides a wide range of machine learning algorithms for various tasks, including classification, regression, clustering, and collaborative filtering. Here are some of the key machine learning algorithms available in Apache Spark's MLlib:

- **Supervised learning**:

  - **Linear regression**: Used to predict a continuous variable based on one or more predictor features

  - **Logistic regression**: Suitable for binary classification problems, predicting the probability of an event occurring

- **Decision trees**: Builds a tree-like model for predicting outcomes based on input features, and can be used for both classification and regression
- **Random forest**: An ensemble method that builds multiple decision trees and combines their predictions for improved accuracy and robustness
- **Gradient-Boosted Trees (GBTs)**: Builds a series of weak decision trees and combines them to create a strong predictive model
- **Naive Bayes**: A probabilistic algorithm commonly used for classification tasks, especially in text classification
- **SVMs**: Suitable for classification and regression tasks, focusing on finding the optimal hyperplane for separation

- **Unsupervised learning**:

  - **K-means**: A clustering algorithm that partitions data into k clusters, based on similarity
  - **Gaussian Mixture Model (GMM)**: A probabilistic model that represents a mixture of Gaussian distributions; used for clustering
  - **PCA**: Reduces the dimensionality of data while retaining as much variability as possible; often used for feature extraction

- **Collaborative filtering**:

  - **Alternating Least Squares (ALS)**: A matrix factorization algorithm, commonly used for collaborative filtering in recommendation systems

- **Feature transformation and selection**:

  - **Word2Vec**: Converts words into vectors, capturing semantic relationships between words. Often used in NLP tasks
  - **Term Frequency-Inverse Document Frequency (TF-IDF)**: TF-IDF is a feature extraction technique commonly used in text mining

The utilities available in Spark include the following:

- **Pipeline**: Allows users to define a sequence of data processing and machine learning stages in a declarative manner
- **Cross-validation**: Supports model selection and hyperparameter tuning by providing tools for training and evaluating models on different subsets of data

These are just some examples of the machine learning algorithms available in Apache Spark's MLlib. The library continuously evolves; newer versions may introduce additional algorithms and improvements. Additionally, Spark's compatibility with other machine learning libraries, such as TensorFlow and scikit-learn, allows users to leverage a broader range of algorithms within Spark environments.

Apache Spark use cases

Apache Spark is widely adopted across various industries and is used by a diverse range of organizations for machine learning tasks. Here are some major users and industries leveraging Apache Spark for machine learning:

- **Technology companies**: Leading technology companies, including those in the fields of cloud computing, data analytics, and artificial intelligence, often use Apache Spark for large-scale machine learning tasks. Companies like Google, Amazon, Microsoft, and IBM integrate Spark into their platforms and services.

- **Financial services**: Banking and financial institutions use Apache Spark for tasks such as fraud detection, risk assessment, customer segmentation, and algorithmic trading. The ability to process large volumes of financial data in real-time makes Spark a valuable tool in this industry.

- **Healthcare and life sciences**: Organizations in the healthcare and life sciences sectors utilize Apache Spark for tasks such as genomics analysis, drug discovery, patient data analytics, and personalized medicine. Spark's ability to handle diverse data types and large datasets is beneficial in these applications.

- **Retail and e-commerce**: Retailers and e-commerce companies leverage Apache Spark for recommendation systems, customer segmentation, demand forecasting, and supply chain optimization. Spark's machine learning capabilities help these businesses extract valuable insights from customer and transaction data.

- **Manufacturing and Industry 4.0**: Manufacturing companies adopt Apache Spark for predictive maintenance, quality control, supply chain optimization, and sensor data analytics. Spark's ability to handle streaming data is particularly valuable in scenarios involving **Internet of Things (IoT)** devices.

- **Energy and utilities**: Energy companies use Apache Spark for tasks such as predictive equipment maintenance, energy consumption forecasting, and grid optimization. Spark's ability to process and analyze time-series data is beneficial in this sector.

Next, we discuss how to install and set up Apache Spark.

Setting up Apache Spark

Setting up Apache Spark for local development involves installing Spark on your machine and configuring it to run in a standalone mode. Here are the general steps to set up Apache Spark for local development:

> **Note**
>
> The following instructions assume that you have Java installed on your machine, which Apache Spark requires.

1. **Download Apache Spark**:

    I.   Visit the official Apache Spark website: `https://spark.apache.org/`.

    II.   Go to the **Download** section.

    III.   Choose the Spark version you want to download.

    IV.   Select the package type. For local development, you can choose the **Pre-built for Apache Hadoop** option.

    V.   Download the tarball (`.tgz`) or ZIP file containing Spark.

2. **Extract the Spark archive**:

    I.   Navigate to the directory where you downloaded the Spark archive.

    II.   Extract the contents of the archive, using a tool like `tar` or a graphical tool if you downloaded a ZIP file:

    ```
    tar -xvf spark-3.x.x-bin-hadoop3.x.tgz
    ```

3. **Configure the environment variables**:

    I.   Open your shell profile configuration file. For example, if you are using Bash, edit your `bashrc` file using the following command:

    ```
    ~/.bashrc
    ```

    Alternatively, use this:

    ```
    ~/.bash_profile
    nano ~/.bashrc
    ```

    II.   Add the following lines to set the environment variables:

    ```
    export SPARK_HOME=/path/to/extracted/spark-3.x.x-bin-hadoop3.x
    exportPATH=$PATH:$SPARK_HOME/bin
    ```

III. Replace `/path/to/extracted/spark-3.x.x-bin-hadoop3.x` with the actual path where you extracted Spark.

IV. Save the file and exit the text editor.

V. Run the following command to apply the changes:

```
source ~/.bashrc
```

4. **Start Spark in standalone mode**:

I. Navigate to the Spark directory:

```
cd $SPARK_HOME
```

II. Start the Spark master:

```
./sbin/start-master.sh
```

III. Start a Spark worker:

```
./sbin/start-worker.sh spark://localhost:7077
```

5. **Access the Spark Web UI**:

I. Open a web browser and go to `http://localhost:8080/` to view the Spark web UI. The master and worker should be listed.

6. **Test the Spark installation**:

I. Open a new terminal window.

II. Navigate to the Spark directory:

```
cd $SPARK_HOME
```

III. Run a simple example to test the installation:

```
./bin/run-example SparkPi
```

This command calculates an approximation of Pi using Spark.

7. **Stop Spark**:

I. Stop the Spark worker and master:

```
./sbin/stop-worker.sh
./sbin/stop-master.sh
```

In addition to setting up Spark using the previous method, here are other ways to run PySpark:

- **Using PyPI**:

```
pip install pyspark
pyspark
```

- **Using a pre-built docker image**:

```
docker run -it --rm spark:python3 /opt/spark/bin/pyspark
```

> **Note**
> The coding examples in this book use Apache Spark version 3.5.1.

This concludes the steps involved in setting up Spark.

Summary

As we conclude this introductory chapter, it is evident that Apache Spark has emerged as a transformative force in the world of big data processing. Its in-memory computing, unified processing engine, and ease of use have positioned it as a go-to solution for organizations grappling with the challenges of large-scale data analysis. Apache Spark has also become a popular platform for large-scale data engineering and machine learning.

In this chapter, we learned about the basics of machine learning, the different types of learning algorithms, the components of Spark, and its benefits in machine learning.

In the subsequent chapters, we will delve deeper into each component of Apache Spark, exploring practical applications and providing hands-on examples to illustrate its capabilities. In the next chapter, we will learn about the various data processing techniques in Apache Spark.

2

Data Processing with Spark

Data processing is an essential step in data analysis and machine learning, as it involves transforming, cleaning, and integrating raw data into a suitable format for further processing. In this chapter, we will introduce the basic concepts and principles of data processing, providing some practical examples and use cases of data preprocessing with Apache Spark.

In this chapter, we will cover the following topics:

- Understanding data preprocessing
- Ingesting data
- Cleaning and transforming data
- Aggregating data
- Windowing in Spark
- Data joining

By the end of this chapter, you will know the different data processing techniques using Spark.

Technical requirements

You can find the code files for this chapter on GitHub at `https://github.com/PacktPublishing/Apache-Spark-for-Machine-Learning/tree/main/Chapter02`.

Understanding data preprocessing

Data preprocessing is a fundamental step in the data mining process. It involves a series of operations on raw data to transform it into a suitable format for further analysis and modeling, particularly in machine learning and AI applications. The key goal of data preprocessing is to enhance the quality, reliability, and efficiency of weather prediction based on historical data.

The main steps in data processing include the following:

1. **Data collection**: The first step involves extracting raw data from various sources. Data can be in multiple formats and can include numbers, text, and images.

2. **Data preparation**: This step involves preparing the data for processing. It might include sorting the data and organizing it into tables, databases, or files.

3. **Data cleaning**: Data cleaning involves filling in the missing values through data imputation, where missing values are replaced with substituted values such as mean, median, and max. It may also involve removing outliers and correcting erroneous, incomplete, irrelevant, or duplicate parts of the data. This step is crucial for maintaining the quality of data.

4. **Data transformation**: This involves transforming or converting data into a suitable format for analysis. It can also involve normalizing or scaling the data, aggregating data points, or converting data types.

5. **Data integration**: If data is gathered from multiple sources, it may need to be combined or integrated into a single coherent dataset. This can involve merging databases, tables, or files and aligning different data formats.

6. **Data encoding**: This includes converting categorical data into a numerical format so that it can be used by machine learning algorithms. Common techniques include label encoding and one-hot encoding.

7. **Feature selection and engineering**: This involves identifying the most relevant features for the analysis or model and creating new features from the existing data, improving the model's performance.

Data preprocessing can improve the performance and scalability of machine learning models by reducing the complexity and size of data, selecting the most relevant features, and handling errors and inconsistencies. Data preprocessing can differentiate between a successful and a failed machine learning project. Therefore, applying appropriate data preprocessing techniques and tools to data is very important before feeding it to machine learning models.

One example of a failed machine project due to a lack of data preprocessing is the **IBM Watson for Oncology** project. The data was not properly cleaned and validated, as it contained errors, inconsistencies, and outdated information. The data was also not integrated and standardized, as it came from different sources and formats, such as medical records, clinical trials, and research papers.

Now, let's see how to ingest data for preprocessing.

Ingesting data

Data ingestion is the process of importing and loading data into a system, such as a database, a data warehouse, or a data lake. Data ingestion can be done manually or automatically, using various tools and techniques. Data ingestion is the first step in data analysis and machine learning, as it prepares the data for further processing and usage.

Apache Spark is a powerful, distributed data processing system that can read from a wide variety of data sources. Its ability to integrate with many diverse types of data storage systems is one of the reasons for its popularity in big data processing and analytics. Here are some of the key data sources from which Apache Spark can ingest data.

Filesystems

Let's explore an example of data ingestion from **Hadoop Distributed File System** (**HDFS**). Here is a sample code snippet to read data from HDFS:

```
from pyspark.sql import SparkSession
spark = SparkSession.builder \
    .appName("HDFS Read Example") \
    .getOrCreate()

# Define the HDFS path
hdfs_path = "hdfs://namenode:port/path/to/your/file.csv"

# Read the CSV file into a DataFrame
df = spark.read.csv(hdfs_path, header=True, inferSchema=True)
# Show the DataFrame
df.show()
```

Amazon S3

Now, let's look at an example of data ingestion from Amazon S3. To read data from an S3 bucket, you can use this code snippet:

```
from pyspark.sql import SparkSession

# Initialize a SparkSession
spark = SparkSession.builder \
    .appName("Read Data from S3") \
    .config("spark.hadoop.fs.s3a.access.key", "YOUR_ACCESS_KEY") \
    .config("spark.hadoop.fs.s3a.secret.key", "YOUR_SECRET_KEY") \
    .config("spark.hadoop.fs.s3a.impl", \
        "org.apache.hadoop.fs.s3a.S3AFileSystem") \
    .getOrCreate()

# Replace 's3a://your-bucket-name/your-data-file' with your S3 bucket
and file
s3_file_path = "s3a://your-bucket-name/your-data-file"

# Read the data into a DataFrame
```

```
df = spark.read.csv(s3_file_path, header=True, inferSchema=True)

# Show the DataFrame
df.show()

# Stop the SparkSession
spark.stop()
```

Azure Blob Storage

Next, we will see an example of data ingestion from Azure Blob Storage. Here's a sample code snippet to read data from Azure Blob Storage:

```
from pyspark.sql import SparkSession
 # Initialize a SparkSession
spark = SparkSession.builder \
    .appName("Read Data from Azure Blob Storage") \
    .config("fs.azure", \
        "org.apache.hadoop.fs.azure.NativeAzureFileSystem") \
    .config(
        "fs.azure.account.key.YOUR_STORAGE_ACCOUNT_NAME.blob.core.
windows.net", \
        "YOUR_STORAGE_ACCOUNT_ACCESS_KEY") \
    .getOrCreate()

# Replace with your Azure Storage account details
storage_account_name = "YOUR_STORAGE_ACCOUNT_NAME"
container_name = "YOUR_CONTAINER_NAME"
file_path = "YOUR_FILE_PATH"  # e.g., "folder/data.csv"
# Construct the full Azure Blob Storage path
blob_file_path = f"wasbs://{container_name}@{storage_account_name} \
                . blob.core.windows.net /{file_path}"
# Read the data into a DataFrame
# Replace 'spark.read.csv' with the appropriate method for your data
format
# Use "spark.read.parquet" for parquet files
# Use "spark.read.json" for JSON files
df = spark.read.csv(blob_file_path, header=True, inferSchema=True)

# Show the DataFrame
df.show()

# Stop the SparkSession
spark.stop()
```

Relational databases

We'll also see an example of data ingestion from relational databases.

The following databases can be used to ingest data in Apache Spark:

- MySQL
- PostgreSQL
- Oracle
- SQL Server

To read data from a PostgreSQL database, you can try this code snippet:

```python
from pyspark.sql import SparkSession
# Initialize a SparkSession
spark = SparkSession.builder \
    .appName("Read Data from PostgreSQL") \
    .getOrCreate()

# PostgreSQL database details
jdbc_url = "jdbc:postgresql://hostname:port/database_name"
properties = {
    "user": "username",
    "password": "password",
    "driver": "org.postgresql.Driver"
}

# Name of the table you want to read
table_name = "your_table"

# Read data from PostgreSQL into a DataFrame
df = spark.read.jdbc(
    url=jdbc_url, table=table_name, properties=properties)

# Show the DataFrame
df.show()

# Stop the SparkSession
spark.stop()
```

NoSQL databases

Let's review an example of data ingestion from NoSQL databases.

The following NoSQL databases can be employed for data ingestion:

- Cassandra

- HBase

- MongoDB

Here is a sample code snippet to read data from a Cassandra NoSQL database:

```python
from pyspark.sql import SparkSession

# Configure the Spark session to use the Spark Cassandra Connector
# Replace with the correct version
connector_package = "com.datastax.spark:spark-cassandra-
connector_2.12:3.0.0"

spark = SparkSession.builder \
    .appName("PySpark Cassandra Example") \
    .config("spark.jars.packages", connector_package) \
    .config("spark.cassandra.connection.host", "cassandra_host") \
    .config("spark.cassandra.connection.port", "cassandra_port") \
    .config("spark.cassandra.auth.username", "username") \
    .config("spark.cassandra.auth.password", "password") \
    .getOrCreate()

# Now you can use the Spark session to read from or write to Cassandra
# Example: Reading data from a Cassandra table
keyspace = "your_keyspace"
table = "your_table"

df = spark.read \
    .format("org.apache.spark.sql.cassandra") \
    .options(keyspace=keyspace, table=table) \
    .load()

df.show()

# Remember to stop the Spark session
spark.stop()
```

Additional data sources

In addition to the data ingestion sources discussed earlier, we have some other formats:

- **Big data file formats**:

  - **Parquet**: An efficient, columnar storage format available to any project in the Hadoop ecosystem

  - **Avro**: A row-based storage format known for its efficiency and effectiveness in serializing enormous amounts of data

  - **Optimized Row Columnar (ORC)**: A format that is highly efficient and provides impressive compression ratios

  - **JavaScript Object Notation (JSON)**: A popular format for data interchange

  - **Comma-Separated Values (CSV)**: A simple format for tabular data

- **Data streams**:

  - **Kafka**: A distributed streaming platform for building real-time data pipelines and streaming applications

  - **Flume**: A service for efficiently collecting, aggregating, and moving enormous amounts of log data

  - **Kinesis and Event Hubs**: Scalable and durable real-time data streaming service from cloud service providers

- **Data warehousing solutions**:

  - **Apache Hive**: A data warehouse system for Hadoop, facilitating easy data summarization, querying, and analysis

  - **Amazon Redshift**: A fast, scalable data warehouse service from Amazon

  - **Google Big Query**: Google's serverless, highly scalable, and cost-effective multi-cloud data warehouse

  - **Elasticsearch**: A distributed, RESTful search and analytics engine to address a growing number of use cases

Apache Spark can handle a wide variety of data processing scenarios, everything from batch processing to real-time streaming data. Its ability to connect to so many diverse types of data sources makes it a powerful tool for data engineers and data scientists.

In the next section, let's understand how to clean and transform data.

Cleaning and transforming data

Cleaning and transforming data are crucial steps in the data preprocessing phase, which prepares raw data for analysis and modeling. Here's what each step involves:

- Data cleaning involves identifying and correcting errors, data corruption, inconsistencies, and inaccuracies in data. This process is essential for ensuring the quality and reliability of data before it's used for analysis.

- Data transformation involves converting data from its original format or structure into a format that's more suitable for analysis. This step is vital for making the data compatible with various analysis tools and techniques.

PySpark provides various functions and methods to clean and transform data efficiently.

Data cleaning

Data cleaning is a critical step in the ML pipeline because it directly impacts the performance and reliability of machine learning models. By ensuring that data is clean and well-organized, you can build models that are more accurate, efficient, and interpretable.

Key tasks in data cleaning include the following:

- **Removing duplicates**: Eliminating repeated entries to ensure that each data point is unique

- **Handling missing values**: Deciding how to deal with gaps in data, such as filling them with a calculated value (mean, median, mode), interpolation, or removing the affected records entirely

- **Correcting errors**: Identifying and fixing mistakes in data, which could be typographical errors, mislabeled categories, or incorrect measurements

- **Standardizing formats**: Ensuring consistency in units, date formats, and categorical labels to avoid discrepancies that can distort analysis

In the next subsections, we will explore some common data cleaning operations in PySpark.

Removing duplicates

To remove duplicate rows from a DataFrame or a subset of columns, you can use the `dropDuplicates()` method:

```
import pyspark
from pyspark.sql import SparkSession
spark = SparkSession \
    .builder \
    .appName("Python Spark SQL basic example") \
    .getOrCreate()
```

```
df = spark.createDataFrame(
    [('tony', 25), ('tony', 25), ('mike', 40)], ["name", "age"])
df.dropDuplicates().show()
```

We will get the following output:

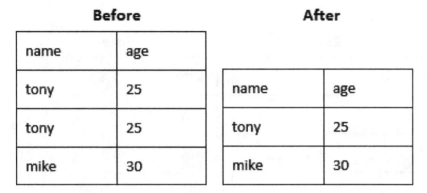

Figure 2.1 – Duplicate record removal

Handling missing values

You can drop rows that have missing values in any column using dropna():

```
# Drops rows with any null values
df = df.dropna(how='any')
```

You will get the following output:

Before

name	age
tony	None
tony	25
mike	30

After

name	age
tony	25
mike	30

Figure 2.2 – Dropping missing value rows

Filling in missing values

We will learn how to fill in the missing values through an example.

Use `fillna()` to replace null values with a specified value:

```
df = spark.createDataFrame([('tony', None), ('tony', 25),
    ('mike', 40)], ["name", "age"])
df.na.fill({'age': 55}).show()
```

Here's the result:

Before

name	age
tony	null
tony	25
mike	30

After

name	age
tony	55
tony	25
mike	30

Figure 2.3 – An overview of filling in missing values

Filtering data

You can filter out rows based on a condition, using the `filter()` or `where()` method:

```
df = df.filter(df['age'] > 20)
```

Here's the updated result:

Before

name	age
tony	15
tony	25
mike	30

After

name	age
tony	25
mike	30

Figure 2.4 – An overview of the filter operation

Now, let's shift our attention to data transformation.

Data transformation

Data transformation is a critical aspect of data processing in machine learning workflows. Apache Spark provides robust functionalities to efficiently perform various data transformations.

Key tasks in data transformation include the following:

- **Changing data types**: This operation involves converting a column from one data type to another, such as from a string to an integer
- **Renaming columns**: This transformation involves assigning new names to columns, which can help to maintain consistency and clarity in your dataset
- **Trimming strings**: This process involves removing leading and trailing spaces from string values in a column
- **Sorting data**: This transformation involves arranging the rows of a dataset based on the values of one or more columns, either in ascending or descending order

Let's review each operation in detail.

Transforming data types

To change the data type of a column, use `withColumn()` along with `cast()`:

```
df = spark.createDataFrame(
    [('tony', 35), ('tony', 25), ('mike', 40)], ["name", "age"])
df.withColumn("age", df["age"].cast("integer")).show()
```

Renaming columns

Next, we will learn how to rename a column through an example. You can rename a DataFrame column using `withColumnRenamed()`:

```
df = df.withColumnRenamed("name", "new_name")
```

We will get the following result:

Before		After	
name	age	New_name	age
tony	15	tony	15
tony	25	tony	25
mike	30	mike	30

Figure 2.5 – An overview of the renaming column operation

Trimming strings

To remove leading and trailing spaces from string columns, use `trim()`:

```
from pyspark.sql.functions import trim
df = df.withColumn("new_column"_name", trim(df["name"]))
```

Here's the result:

Before

name	age
"tony"	15
"tony"	25
" mike"	30

After

name	age	Name_new
"tony"	15	"tony"
"tony"	25	"tony"
" mike"	30	" mike"

Figure 2.6 – An overview of trimming strings

Sorting data

Sort data using `orderBy()`:

```
df = df.orderBy(df["age"].desc())
```

Here is the output:

Before

name	age
tony	25
tony	15
mike	30

After

name	age
tony	15
tony	25
mike	30

Figure 2.7 – An overview of the sorting operation

In the next section, we will learn about data aggregation.

Aggregating data

Data aggregation, in the context of artificial intelligence and machine learning, refers to the process of transforming raw data into a summary or a simplified representation that is more suitable for analysis or model training.

Aggregation is a critical step in data preprocessing and analysis for several reasons:

- **Reduction of complexity**: Aggregation helps to transform several rows of data into fewer rows of data by summarizing detailed data. For example, individual sales transactions could be aggregated to show total sales per day, month, or year. This reduction in complexity can make it easier for algorithms to identify patterns and trends.

- **Improving signal-to-noise ratio**: By summarizing or averaging data, aggregation can enhance the underlying signal (relevant information) while reducing noise (irrelevant information). This can lead to better model performance, especially in noisy datasets.

- **Dimensionality reduction**: Aggregation can reduce the number of features in a dataset (dimensionality reduction), which is beneficial for many machine learning algorithms. It helps to mitigate issues such as overfitting and can improve the efficiency and performance of models.

- **Handling time-series data**: In time-series analysis, aggregation is often used to resample data into different time frames (for example, converting five-second interval data into minute-level summaries). This is crucial for models that require data at specific intervals or need a simplified view of trends over time.

- **Enhancing data quality**: Aggregation can help to deal with missing or incomplete data. By aggregating data, the impact of missing values can be lessened, especially if the aggregation process involves calculations such as averages or sums.

- **Facilitating feature engineering**: Aggregated data can be used to create new features that provide more insight than the original raw data. For instance, aggregating customer purchase behavior over time can create features that represent long-term trends or customer loyalty.

- **Resource efficiency**: Working with aggregated data can be computationally less intensive than working with raw, detailed data. This can be particularly important when dealing with exceptionally large datasets (big data).

In practice, the choice of aggregation method and the level of aggregation depend on the specific goals of the AI/ML project, the nature of the data, and the type of patterns or insights that a project aims to uncover. Common aggregation functions include sum, average, count, max, min, and median. Advanced aggregation might involve more complex statistical methods or custom algorithms tailored to specific data characteristics or business requirements.

There are two types of aggregations as we will discuss next.

Basic aggregations

Basic aggregation in data analysis involves applying simple statistical or mathematical operations to summarize data. These operations reduce data from many values to a single value, which provides a simplified overview or key insights into the dataset. Here are some of the most common types of basic aggregation functions:

- `max()`: Finds the maximum value in a column

- `min()`: Finds the minimum value in a column

- `mount()`: Counts the number of rows in a table

- `sum()`: Adds up the values in a column

- `mean()`: Calculates the average of values in a column

Let's look at the dataset containing four columns as shown in *Figure 2.8*. Since the aggregate functions can be applied only on numerical columns, applying the various aggregate functions on the column "amount" yields output, as shown in the right-hand section of the following diagram:

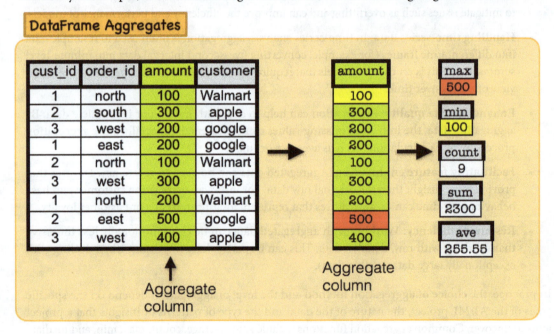

Figure 2.8 – An example of basic aggregation

The following code creates a Spark DataFrame, df, containing four columns, named emp_id, region, sales, and customer:

```
df = spark.createDataFrame(
    [(1,"north",100,"walmart"),(2,"south",300,"apple"),
    (3,"west",200,"google"),(1,"east",200,"google"),
    (2,"north",100,"walmart"),(3,"west",300,"apple"),
    (1,"north",200,"walmart"),(2,"east",500,"google"),
    (3,"west",400,"apple"),],["emp_id","region","sales","customer"])
```

Now that we have created a DataFrame, we can apply the aggregate function, sum, on the sales column:

```
df.agg({"sales": "sum"}).show()
```

The sum function adds all the values in the column to give an output value of 2300:

```
+----------+
|sum(sales)|
+----------+
|      2300|
+----------+
```

The following code calculates the minimum value in the sales column:

```
df.agg({"sales": "min"}).show()
```

Here is the output:

```
+----------+
|min(sales)|
+----------+
|       100|
+----------+
```

The following code calculates the maximum value in the sales column:

```
df.agg({"sales": "max"}).show()
```

Here is the output:

```
+----------+
|max(sales)|
+----------+
|       500|
+----------+
```

The following code calculates the count of number of rows in the dataset:

```
df.agg({"sales": "count"}).show()
```

Here is the output:

```
+------------+
|count(sales)|
+------------+
|           9|
+------------+
```

The following code calculates the average of all the values in the `sales` column:

```
df.agg({"sales": "mean"}).show()
```

Here is the output:

```
+------------------+
|        avg(sales)|
+------------------+
|255.55555555555554|
+------------------+
```

The following code shows how to apply two aggregate functions simultaneously:

```
df.agg({"sales": "mean","customer":"count"}).show()
```

Here is the output:

```
+------------------+---------------+
|        avg(sales)|count(customer)|
+------------------+---------------+
|255.55555555555554|              9|
+------------------+---------------+
```

Grouped aggregations

Grouped aggregation in data processing, especially in the context of SQL, data analytics, and programming languages such as Python, refers to the process of performing calculations or statistical operations on a dataset by dividing it into subsets or groups. This is particularly useful when you want to analyze patterns or compute metrics within specific segments of your data. Grouped aggregations are fundamental in data analysis, allowing for detailed and segmented insights into datasets. They transform large, detailed datasets into summarized, actionable information.

Let's see how grouped aggregation works:

1. **Grouping data**: First, the data is divided into groups based on one or more keys. These keys can be columns in a dataset, such as a category, date, or region.

2. **Applying aggregation functions**: Once the data is grouped, various aggregation functions can be applied to each group independently. Common aggregation functions include sum, average (mean), count, maximum, minimum, and standard deviation.

3. **Result**: The result is a new dataset where each group is represented by a single row, and the columns contain the aggregated values.

The following diagram illustrates how you can apply the `GroupBy` function on various columns to group together records, based on a column value. Grouped records are shown in the center section.

Aggregate functions can then be applied to each of these grouped records to get group-specific aggregates, as shown in the right-hand section.

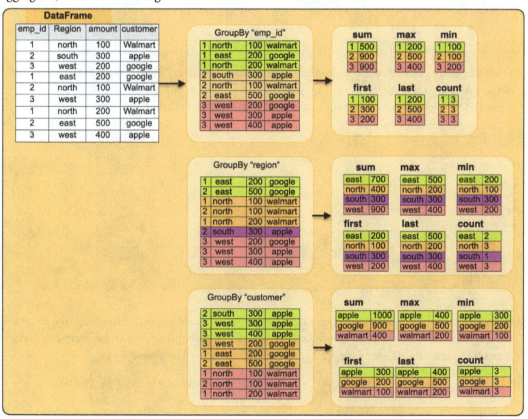

Figure 2.9 – An example of applying aggregate functions to grouped data

Method chaining in PySpark allows you to perform multiple operations on a DataFrame concisely and sequentially. Instead of assigning intermediate DataFrames to variables, you can chain methods together directly.

The following code calculates the total sales achieved by every employee. First, it groups the records in the `emp_id` column for each unique value of `emp_id`. For each group, (that is, for every employee ID), it calculates the sum. Then, the output is sorted by employee ID.

```
df.groupby("emp_id").agg({"sales": "sum"}).orderBy('emp_id').show()
```

Here is the output:

```
+------+----------+
|emp_id|sum(sales)|
+------+----------+
|     1|       500|
|     2|       900|
|     3|       900|
+------+----------+
```

The following code calculates the maximum sales achieved by every employee. It first groups the records in the `emp_id` column for each unique value of `emp_id`. For each group (that is, for every employee ID), it calculates the maximum value. Then, the output is sorted by employee ID.

```
df.groupby("emp_id").agg({"sales": "max"}).orderBy('emp_id').show()
```

Here is the output:

```
+------+----------+
|emp_id|max(sales)|
+------+----------+
|     1|       200|
|     2|       500|
|     3|       400|
+------+----------+
```

The following code outputs the most recent sales achieved by every employee. It first groups the records in the `emp_id` column for each unique value of `emp_id`. For each group (that is, for every employee ID), it reads the last sales in every group. Then, the output is sorted by employee ID.

```
df.groupby("emp_id").agg({"sales": "last"}).orderBy('emp_id').show()
```

Here is the output:

```
+------+-----------+
|emp_id|last(sales)|
+------+-----------+
|     1|        200|
|     2|        500|
|     3|        400|
+------+-----------+
```

The following code outputs the average sales for each region. It first groups the records in the `region` column for each unique value of `region`. For each group (that is, for every region), it calculates the average sales in every group. Then, the output is sorted by region.

```
df.groupby("region").agg({"sales": "sum"}).orderBy('region').show()
```

Here is the output:

```
+------+----------+
|region|sum(sales)|
+------+----------+
|  east|       700|
| north|       400|
| south|       300|
|  west|       900|
+------+----------+
```

The following code outputs the total sales for each customer. It first groups the records in the `customer` column for each unique value of `customer`. For each group (that is, for every customer), it calculates the total sales. Then, the output is sorted by customer.

```
df.groupby("customer").agg({"sales": "sum"}).orderBy('customer').
show()
```

Here is the output:

```
+--------+----------+
|customer|sum(sales)|
+--------+----------+
|   apple|      1000|
|  google|       900|
| walmart|       400|
+--------+----------+
```

In the next section, we will learn about data windowing.

Windowing in Spark

Data windowing is a crucial technique in machine learning and artificial intelligence, particularly when dealing with time series data, sequential data, or any data where temporal or sequential context is important. It splits the large sequential series of data points into smaller sequential series (data windows) to derive insights from each of those data partitions. In these domains, windowing helps in structuring the data appropriately for analysis and model training. Here is how it is applied in ML and AI:

- **Feature engineering for time series data**:
  - **Creating time frames**: Windowing can be used to create fixed-size time frames from continuous time series data. For instance, if you have continuous sensor data, you can create windows of one-minute intervals to analyze the sensor readings in these discrete chunks.
  - **Capturing temporal patterns**: By segmenting data into windows, you can capture temporal patterns, which are crucial for forecasting models such as ARIMA, **Long Short-Term Memory (LSTM)** networks, and **Recurrent Neural Networks (RNNs)**.

- **Handling sequential data**:
  - **Natural Language Processing (NLP)**: In text analysis, windowing techniques are used to process words in the context of their surrounding words, which is crucial for understanding language structure and meaning.
  - **Genomics**: In DNA sequence analysis, windowing helps to analyze specific regions of a genome sequence to identify patterns or anomalies.

- **Signal processing**:
  - **Feature extraction from signals**: Windowing is used to extract features from audio, ECG signals, or other time-dependent signals for classification or analysis tasks.
  - **Reducing edge effects in spectral analysis**: In the Fourier Transform and related spectral analysis techniques, window functions are used to mitigate edge effects, ensuring more accurate frequency domain representations.

- **Anomaly detection**: Windowing can help to identify sudden changes in metrics or patterns over time, which is crucial in anomaly detection systems.

- **Data smoothing and noise reduction**: Windowing techniques are used to apply moving averages or other filters to smooth out data, reducing noise and making underlying patterns more apparent.

- **Real-time data processing**: In streaming data analysis, windowing is used to process data in real time, segmenting continuous streams of data into manageable, discrete chunks for analysis.

- **Window types in ML and AI**:

  - **Sliding windows**: An overlap between successive windows, commonly used in speech and audio processing

  - **Tumbling windows**: Non-overlapping, sequential windows, often used in real-time analytics

  - **Expanding windows**: Start from a fixed point and grow over time, useful in cumulative metrics calculation

Why windowing is required and its examples in Spark

Let's explore some key reasons why windowing is necessary and beneficial:

- **Performing calculations across rows**: Window functions perform calculations across a set of rows related to the current row without collapsing them into a single output row, unlike traditional aggregate functions. This is essential for analyses where you need to retain individual row details.

- **Data partitioning**: Window functions enable you to partition data into groups without altering the row structure of the output. This partitioning allows for comparative and aggregate analysis within subsets of data, such as calculating sums, averages, or other metrics within each group.

- **Ranking and row numbering**: They provide the ability to rank items within a dataset without the need for complex subqueries. Functions such as ROW_NUMBER(), RANK(), and DENSE_RANK() are invaluable for tasks such as ranking sales data by region or department, identifying top-performing products, or assigning unique identifiers to rows based on a specific order.

- **Running totals and moving averages**: Window functions are particularly useful for financial and statistical analyses, such as calculating running totals, moving averages, or cumulative sums. This capability is essential for time series analysis, budgeting, financial reporting, and trend analysis.

- **Accessing previous and next row values**: Functions such as LEAD() and LAG() allow you to access data from preceding or following rows within the same result set. This feature is crucial for comparing current row data with previous or next rows, such as calculating period-over-period changes in sales or stock movements, or detecting sequence patterns.

Let's go through some example problems to understand several groupby operations.

Problem 1

Problem statement: How do we calculate a new column for each group whose row value is equal to the sum of the current row and the previous two rows?

As shown in *Figure 2.10*, the dataset consists of three columns – namely, dept_name, emp_id, and salary. We create a new column, sum. Each row value in this new sum column is a sum of the current row value, plus two previous row values in the salary column.

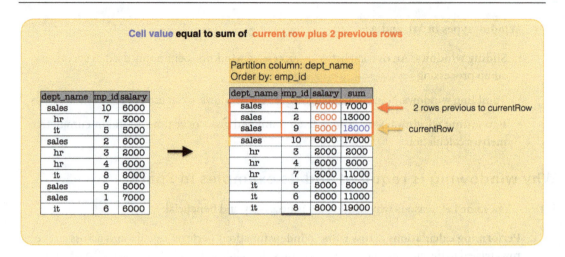

Figure 2.10 – A windowing example

Here's the explanation of how each row's value is calculated from the previous rows:

- The first row value of the sum column = the current row value of salary + the two previous row values:

  7,000 = 7,000 + 0 (note that no previous rows exist, hence 0).

- Similarly, the second row value of the sum column = the current row value of salary + two previous row values:

  13,000 = 6,000 + 7,000 (note that only one previous row exists, hence 7,000).

- Similarly, the third row value of the sum column = the current row value of salary + two previous row values:

  18,000 = 5,000 + 13,000 (note only two previous rows exist, hence 6,000 and 7,000)

Now, let's write the code to perform windowing functions, deriving a new sum column in the dataset.

In *Figure 2.10*, the dataset in the left section is untransformed. We first define a window function, named window. This function partitions the dataset by applying partitionBy to the dept_id column. This action groups the records based on unique values in the dept_id column. It then sorts the emp_id column for each grouped record. Then, the function defines the start and end point of the windows through rowsBetween. The minus symbol denotes preceding rows.

```
from pyspark.sql import functions as func
from pyspark.sql import Window
df = spark.createDataFrame(
    [(1,"north",100,"walmart"),(2,"south",300,"apple"),
```

```
        (3,"west",200,"google"),(1,"east",200,"google"),
        (2,"north",100,"walmart"),(3,"west",300,"apple"),
        (1,"north",200,"walmart"),(2,"east",500,"google"),
        (3,"west",400,"apple"),],
        ["emp_name","region","sales","customer"])
window = Window.partitionBy("dept_id")\
    .orderBy("emp_id").rowsBetween(-2,0)
```

We will now create a new column, named sum, using the aggregate sum function and the window function defined in the previous line of code:

```
df.withColumn("sum",func.sum("salary").over(window)).show()
```

The following output shows the transformed dataset with a new column, sum:

```
+---------+------+------+-----+
|dept_name|emp_id|salary|  sum|
+---------+------+------+-----+
|   sales |    1|  7000| 7000|
|   sales |    2|  6000|13000|
|   sales.|    9|  5000|18000|
|   sales |   10|  6000|17000|
|       hr|    3|  2000| 2000|
|       hr|    4|  6000| 8000|
|       hr|    7|  3000|11000|
|       it|    5|  5000| 5000|
|       it|    6|  6000|11000|
|       it|    8|  8000|19000|
+---------+------+------+-----+
```

Problem 2

Problem statement: How we do calculate a new column for each group whose row value is equal to the sum of the current row and the two following rows?

As shown in *Figure 2.11*, the dataset consists of three columns – namely, dept_id, emp_id, and salary. We create a new column, sum. Each row value in this new sum column is a sum of the current row value plus the two following row values in the salary column.

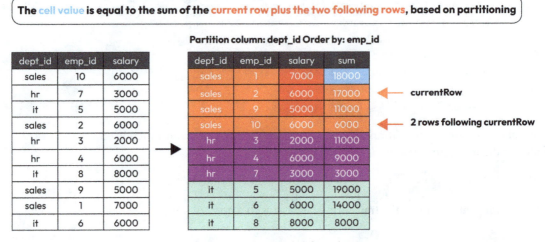

Figure 2.11 – An example of windowing

Here's the explanation of how each row's value is calculated from the previous rows:

- The first row value of the sum column = the current row value of salary + the two following row values:

  18,000 = 7,000 + 6,000 + 5,000

- Similarly, the second row value of the sum column = the current row value of salary + the two following row values:

  17,000 = 6,000 + 5,000 + 6,000

- Similarly, the last row value of the sum column = the current row value of salary + the two following row values:

  8,000 = 8,000 + 0 (note that no following row exists, hence 0)

In *Figure 2.11*, the dataset in the left-hand section is untransformed. We first define the window function. This function partitions the dataset by applying partitionBy to the column. This action groups the records based on unique values in the dept_id column. It then sorts the emp_id column for each grouped record. Then, the function defines the start and end point of the windows through rowsBetween.

```
window = Window.partitionBy("dept_id")\
    .orderBy("emp_id").rowsBetween(0, 2)
```

We will now create a new column, named sum, using the aggregate sum function and the window function defined in the previous line of code:

```
df.withColumn("sum",func.sum("salary").over(window)).show()
```

The following output shows the transformed dataset with a new column, sum:

```
+--------+------+------+-----+
|dept_id|emp_id|salary|  sum|
+--------+------+------+-----+
|  sales|     1|  7000|18000|
|  sales|     2|  6000|17000|
|  sales|     9|  5000|11000|
|  sales|    10|  6000| 6000|
|     hr|     3|  2000|11000|
|     hr|     4|  6000| 9000|
|     hr|     7|  3000| 3000|
|     it|     5|  5000|19000|
|     it|     6|  6000|14000|
|     it|     8|  8000| 8000|
+--------+------+------+-----+
```

In the next section, we will see how to calculate the lag.

How to calculate the lag

The LAG function in SQL is a type of window function that allows you to access data from preceding rows in the same dataset, without the need for a self-join. It's particularly useful for comparing the current row values with the values of preceding rows within the same dataset.

The use cases are as follows:

- Calculating differences between rows
- Identifying trends
- Calculating time intervals between rows

As shown in *Figure 2.12*, the dataset consists of three columns – namely, dept_id, emp_id, and salary. We will create a new column, prev. Each row value in this new prev column is a preceding row value.

Lag – returning a value before the current row

Partition column: dept_id
Order by: salary_desc

dept_id	emp_id	salary
sales	10	6000
hr	7	3000
it	5	5000
sales	2	6000
hr	3	2000
hr	4	6000
it	8	8000
sales	9	5000
sales	1	7000
it	6	6000

dept_id	emp_id	salary	prev
sales	1	7000	NaN
sales	10	6000	7000.0
sales	2	6000	6000.0
sales	9	5000	6000.0
hr	4	6000	NaN
hr	7	3000	6000.0
hr	3	2000	3000.0
it	8	8000	NaN
it	6	6000	8000.0
it	5	5000	6000.0

Figure 2.12 – An example of LAG

Let's walk through how the LAG values are calculated:

- The first row value of the prev column = the previous row value of salary:

  7,000 = NaN (note that no previous rows exist, hence NaN)

- Similarly, the second row value of the sum column = the previous row value of salary:

  7,000 = 7,000 (note that the previous row value is 7,000)

The following code snippet creates the window and then calculates the lag over it:

```
window = Window.partitionBy("dept_id") \
    .orderBy(func.col("salary").desc())
df.withColumn("previousrow_salary", \
    func.lag('salary',1).over(window)).show()
```

Here is the output:

```
+-------+------+------+------------------+
|dept_id|emp_id|salary|previousrow_salary|
+-------+------+------+------------------+
|  sales|     1|  7000|              null|
|  sales|    10|  6000|              7000|
|  sales|     2|  6000|              6000|
```

```
|   sales|      9|  5000|              6000|
|      hr|      4|  6000|              null|
|      hr|      7|  3000|              6000|
|      hr|      3|  2000|              3000|
|      it|      8|  8000|              null|
|      it|      6|  6000|              8000|
|      it|      5|  5000|              6000|
+-------+------+------+------------------+
```

In the next section, we will look at another important data operation – namely, joining.

Data joining

Data joining is a crucial step in the data preparation phase for machine learning and artificial intelligence projects. Its importance is due to several key factors:

- **A comprehensive dataset**: Machine learning models require comprehensive datasets to learn from. Joining data from multiple data sources enables you to create a complete and more comprehensive dataset. This integrated dataset can provide a broader picture of the problem domain, improving a model's ability to make accurate predictions or decisions.

- **Feature enrichment**: Joining datasets can introduce additional features (variables) that can be critical for a model's performance. These new features, derived from combining different data sources, can provide more insights and improve the model's ability to learn complex patterns.

- **Enhanced contextual understanding**: Combining datasets can add context to the data that a machine learning model uses. For instance, joining customer transaction data with customer demographic information can help a model better understand purchasing behaviors.

- **Data quality improvement**: Joining data can also help identify inconsistencies, duplicates, or errors across different data sources, which are essential to address in data preprocessing. Clean and high-quality data is essential for building effective ML models.

- **Time-series data enhancement**: In time-series analysis, joining data from different sources can provide additional temporal context, which can be crucial for forecasting and anomaly detection tasks.

- **Supporting domain-specific requirements**: In specialized domains such as genomics, robotics, or climatology, joining various datasets is often necessary to meet the specific requirements of the analysis or to adhere to domain-specific standards.

In summary, data joining is an essential process in preparing data for machine learning and AI. It helps to enrich, clean, and enhance datasets, thereby directly impacting the performance and accuracy of the resulting models.

Types of data joins

There are several types of data joins used to combine records from two or more tables, based on related columns. Understanding these different join types is crucial for effectively manipulating and analyzing data. *Figure 2.13* shows the most common types of joins supported in Apache Spark:

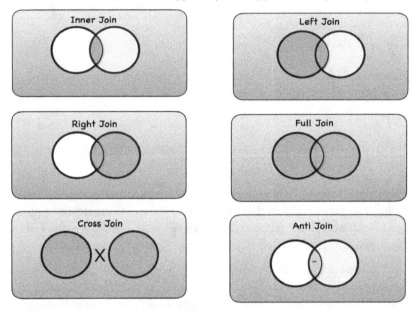

Figure 2.13 – The various data join types

Here's a quick overview of these join types:

- **Inner join**: Returns rows that have matching values in both tables. If a row in either of the tables does not have a match, it is not included in the result set.

- **Left join**: Returns all rows from the left table and the matched rows from the right table. If there is no match, the result is null on the right side.

- **Right join**: Returns all rows from the right table and the matched rows from the left table. If there is no match, the result is null on the left side.

- **Full join**: Returns all rows when there is a match in either the left or right table. This join combines the results of both the left and right outer joins.

- **Cross join**: Returns the Cartesian product of the two tables, which means it combines each row of the first table with each row of the second table. This type of join does not require a condition and can result in an exceptionally substantial number of rows in the output.

- **Anti join**: Returns rows from the first table where no matches are found in the second table. It is the opposite of an inner join.

Each type of join serves a different purpose and is chosen based on the data requirements of the specific task at hand. Understanding how these joins work is fundamental to relational database operations and data analysis tasks.

Let's now discuss the joins shown in *Figure 2.13* in detail.

The inner join

The inner join is the default join in Spark SQL. It selects rows that have matching values in both tables.

Figure 2.14 explains the inner join operation. As seen in the figure, there are two tables, named `Left_Table` and `Right_table`. The `cust_id` column exists in both the tables and, hence, can be used as a join column.

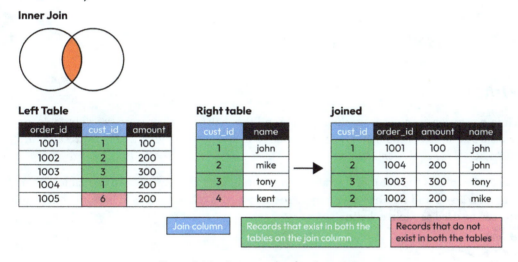

Figure 2.14 – An example of an inner join

The following code creates a Spark DataFrame, `df_left`, containing three columns, named `order_id`, `customer_id`, and `amount`:

```
df_left = spark.createDataFrame(
    [(1001,1,100),(1002,2,200),
    (1003,3,300),(1004,1,200),(1005,6,200)],
    ["order_id","customer_id","amount"])
df_left.show()
```

The following code creates a Spark DataFrame, `df_right`, containing two columns named `customer_id` and name:

```
df_right = spark.createDataFrame(
    [(1,"john"), (2,"mike"),(3,"tony"),(4,"kent")],
```

```
    ["customer_id","name"])
df_right.show()
```

The following code joins two tables using the `cust_id` column:

```
df_left.join(df_right,on="cust_id",how="inner").show()
```

We get the following output:

```
+-------+--------+------+-----+
|cust_id|order_id|amount|name |
+-------+--------+------+-----+
|   1   |  1001  |  100 | john|
|   1   |  1004  |  200 | john|
|   2   |  1002  |  200 | mike|
|   3   |  1003  |  300 | tony|
+-------+--------+------+-----+
```

The left join

A left join returns all values from the left table and the matched values from the right table, or it appends NULL if there is no match. It is also referred to as a left outer join. LEFT JOIN and LEFT OUTER JOIN are equivalent.

Figure 2.15 explains the left join operation. In the following figure, there are two tables, named Left_Table-1 and Right_table-1. The cust_id column exists in both the tables and, hence, can be used as a join column.

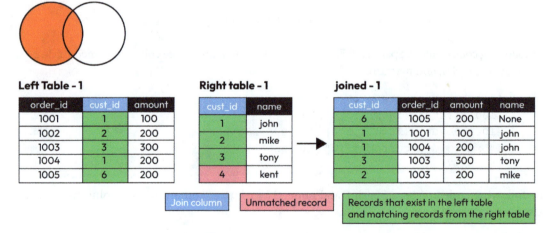

Figure 2.15 – An example of a left join

The following code joins two tables using the `cust_id` column:

```
df_left.join(df_right,on="cust_id",how="left").show()
df_left.join(df_right,on="cust_id",how="left_outer").show()
df_left.join(df_right,on="cust_id",how="leftouter").show()
```

Here is the output:

```
+-------+--------+------+-----+
|cust_id|order_id|amount|name |
+-------+--------+------+-----+
|   6   |  1005  |  200 | NULL|
|   1   |  1001  |  100 | john|
|   1   |  1004  |  200 | john|
|   3   |  1003  |  300 | tony|
|   2   |  1002  |  200 | mike|
+-------+--------+------+-----+
```

The right join

The right join fetches data if it is present in the right table, even if there are no matching records in the left table.

Figure 2.16 explains the right join operation. In the following figure, there are two tables, named `Left_Table-1` and `Right_table-1`. The `cust_id` column exists in both the tables and hence can be used as Join column.

Right Join

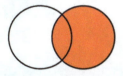

Left Table - 1

order_id	cust_id	amount
1001	1	100
1002	2	200
1003	3	300
1004	1	200
1005	6	200

Right table - 1

cust_id	name
1	john
2	mike
3	tony
4	kent

joined - 1

cust_id	order_id	amount	name
1	1001	100	john
1	1004	200	john
3	1003	300	tony
2	1002	200	mike
4	NaN	NaN	kent

Join column	Unmatched record	Records that exist in the right table and matching records from the left table

Figure 2.16 – An example of a right join

The following code joins two tables using the `cust_id` column:

```
df_left.join(df_right,on="cust_id",how="right").show()
df_left.join(df_right,on="cust_id",how="right_outer").show()
df_left.join(df_right,on="cust_id",how="rightouter").show()
```

Here is the output:

```
+-------+--------+------+-----+
|cust_id|order_id|amount|name |
+-------+--------+------+-----+
|   1   |  1001  | 100  | john|
|   1   |  1004  | 200  | john|
|   3   |  1003  | 300  | tony|
|   2   |  1002  | 200  | mike|
|   4   |  NaN   | NaN  | kent|
+-------+--------+------+-----+
```

The full join

The full join fetches data if it is present in either of the two tables.

Figure 2.17 explains the FULL join operation. In the following figure, there are two tables, named `Left_Table-1` and `Right_table-1`. The `cust_id` column exists in both the tables and, hence, can be used as a join column.

Full Join

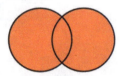

Left Table - 1

order_id	cust_id	amount
1001	1	100
1002	2	200
1003	3	300
1004	1	200
1005	6	200

Right table - 1

cust_id	name
1	john
2	mike
3	tony
4	kent

joined - 1

cust_id	order_id	amount	name
6	1005	200	None
1	1001	100	john
1	1004	200	john
3	1003	300	tony
2	1002	200	mike
4	NaN	NaN	kent

Join column	Records that exist in both the right table and left table

Figure 2.17 – An example of a full Join

The following code joins two tables using the `cust_id` column:

```
df_left.join(df_right,on="cust_id",how="full").show()
df_left.join(df_right,on="cust_id",how="fullouter").show()
df_left.join(df_right,on="cust_id",how="full_outer").show()
```

The output matches what is shown in *Figure 2.17*.

The cross join

A cross join returns the Cartesian product of two relations.

The cross join is an extremely expensive data join operation, often requiring large memory. Hence, it needs to be enabled before using it, as shown in the following code:

```
spark.conf.set("spark.sql.crossJoin.enabled", "true")
df_left.crossJoin(df_right).show()
```

This output is shown in *Figure 2.18*, which explains the cross join operation. In the following figure, there are two tables, named `Left Table-1-1` and `Right table-1-1`. The `cust_id` column exists in both the tables and, hence, can be used as a join column.

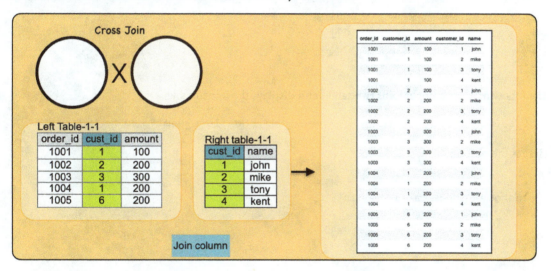

Figure 2.18 – An example of a cross join

The following code demonstrates the joining of two tables using joins such as `semi`, `leftsemi`, and `left_semi`:

```
df_left.join(df_right,on="customer_id",how="semi").show()
```

Here is the output:

```
+-----------+--------+------+
|customer_id|order_id|amount|
+-----------+--------+------+
|          1|    1001|   100|
|          1|    1004|   200|
|          3|    1003|   300|
|          2|    1002|   200|
+-----------+--------+------+
```

The following code snippet joins two tables and displays the joined table:

```
df_left.join(df_right,on="customer_id",how="leftsemi").show()
```

Here is the output:

```
+-----------+--------+------+
|customer_id|order_id|amount|
+-----------+--------+------+
|          1|    1001|   100|
|          1|    1004|   200|
|          3|    1003|   300|
|          2|    1002|   200|
+-----------+--------+------+
```

The following code snippet joins two tables and displays the joined table:

```
df_left.join(df_right,on="customer_id",how="left_semi").show()
```

Here is the output:

```
+-----------+--------+------+
|customer_id|order_id|amount|
+-----------+--------+------+
|          1|    1001|   100|
|          1|    1004|   200|
|          3|    1003|   300|
|          2|    1002|   200|
+-----------+--------+------+
```

Anti join

An anti join returns the rows from one DataFrame that do not have corresponding rows in another DataFrame. It's similar to LEFT OUTER JOIN, where the result is filtered to include only the rows from the left DataFrame for which there is no match in the right DataFrame.

Figure 2.19 explains the anti join operation. In the following figure, there are two tables, named `Left Table-1-2` and `Right table-1-2`. The `cust_id` column exists in both the tables and, hence, can be used as a join column.

Anti Join

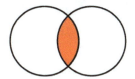

Left Table - 1-2

order_id	cust_id	amount
1001	1	100
1002	2	200
1003	3	300
1004	1	200
1005	6	200

Right table - 1-2

cust_id	name
1	john
2	mike
3	tony
4	kent

Joined Table

cust_id	order_id	amount
1005	6	200

Join column	Records that exist in both the right table and left table

Figure 2.19 – An example of an anti join

The following code demonstrates the joining of two tables using joins such as `anti`, `leftanti`, and `left_anti`:

```
df_left.join(df_right,on="customer_id",how="anti").show()
df_left.join(df_right,on="customer_id",how="leftanti").show()
df_left.join(df_right,on="customer_id",how="left_anti").show()
```

Here's the output showing the result for all three lines:

```
+-----------+--------+------+
|customer_id|order_id|amount|
+-----------+--------+------+
|          6|    1005|   200|
+-----------+--------+------+
```

This concludes our walk-through of several code examples to understand joins.

Summary

In this chapter, we learned about data preprocessing and how Spark supports various data operations, such as ingesting data, cleaning and transforming it, aggregating it, data windowing, and data joining. We learned about the different types of data operations through several coding examples.

In the next chapter, we will learn about feature reengineering.

3

Feature Extraction and Transformation

Feature extractors and transformers are key components in the field of machine learning and data preprocessing. They help to convert raw data into a suitable and more effective format to train machine learning models. Feature extractors simplify data inputs by transforming them into a lower-dimensional space. Conversely, feature transformers involve modifying or transforming the features themselves into a format that's more suitable for machine learning algorithms.

Understanding these concepts is crucial for building efficient and accurate predictive models.

In this chapter, we're going to cover the following main topics:

- Learning about feature extractors
- Working with feature transformers
- Exploring feature selectors

By the end of this chapter, you will know several different techniques to extract features through data transformation.

Technical requirements

You can find the code files for this chapter on GitHub at `https://github.com/PacktPublishing/Apache-Spark-for-Machine-Learning/tree/main/Chapter03`.

Learning about feature extractors

In this section, we will learn what a feature extractor is, why it is very important in machine learning, and its various techniques. A feature extractor in machine learning and data processing is a method or algorithm that automatically identifies and extracts relevant features from raw data, for use in building machine learning models. Feature extraction improves the efficiency and accuracy of subsequent analyses and predictions.

The key aspects of feature extractors

Feature extraction is a critical process in machine learning and data analysis, where the goal is to reduce the number of input variables to those that are most meaningful to the task at hand. It involves transforming raw data into a set of features that are efficient and effective for modeling.

Here are the key aspects of feature extraction:

- **Capturing relevant information**: Effective feature extraction aims to capture the most relevant information from raw data. This involves identifying which features are most informative and representative of the underlying patterns or structures in the data.

- **Data simplification**: By extracting key features, the complexity of the data is reduced, making it easier to work with. This simplification can lead to more efficient data processing and analysis, especially with large datasets.

- **Improving model accuracy and performance**: By focusing on the most important features, feature extraction can lead to improved model accuracy and performance. It helps a model to concentrate on the aspects of data that are most predictive, rather than being distracted by noise or irrelevant information.

- **Handling diverse data types**: Feature extraction techniques vary based on the type of data, such as numerical, categorical, text, or image. Each data type requires different methods to effectively extract features.

- **Contextual and domain knowledge**: Incorporating domain knowledge can significantly enhance the feature extraction process. Understanding the context and specific characteristics of data can guide the selection of appropriate features and extraction techniques.

- **Balancing information loss and efficiency**: There is often a trade-off between information retention and efficiency. While reducing the feature space can make models more efficient, it can also lead to the loss of potentially important information. The challenge is to strike a balance that maximizes efficiency without significantly compromising the quality of information.

- **Facilitating better data visualization and interpretation**: Extracted features can often make data visualization and interpretation more manageable and insightful, as they distill data into its most informative components.

Algorithms for feature extraction

Let's review some of the algorithms used in feature extractors:

- **Text data**:
  - **Bag of words**: Transforms text into a fixed length set of features, representing the frequency of certain words.

- **Term Frequency-Inverse Document Frequency (TF-IDF)**: This is a measure of the importance of a word to a document in a collection of documents. It adjusts the frequency counts of a word depending on how often they appear in an entire dataset, highlighting words that are unique to a particular document.

- **Word embeddings (for example, Word2Vec and GloVe)**: Converts words into high-dimensional vectors that capture semantic relationships between words, allowing a model to understand word meanings based on their context.

- **Image data**:

  - **Histogram**: Used in image processing and computer vision for object detection, this counts occurrences of gradient orientation in localized portions of an image

- **Audio data**:

  - **Mel-frequency cepstral coefficients**: Widely used in speech and audio processing, they represent the short-term power spectrum of a sound

  - **Spectral centroid and spectral roll-off**: Measure the "brightness" and "shape" of the sound spectrum, respectively

  - **Zero crossing rate**: The rate at which a signal changes from positive to negative or back, making it useful for analyzing the noisiness of an audio signal

- **Time series data**:

  - **Fourier transform**: Converts time series data into the frequency domain, useful for identifying cyclic patterns

  - **Statistical features**: Mean, median, standard deviation, and so on, computed over windows of the time series

  - **Autocorrelation**: Measures how the values of the series are related to themselves over different time lags

- **Signal processing**:

  - **Wavelet transform**: Provides a multi-resolution analysis of signals, capturing both frequency and location information

  - **Short-time Fourier transform**: A Fourier transform is applied to short, overlapping windows of a signal, making it useful for non-stationary signal analysis

- **General data processing**:

  - **Principal Component Analysis (PCA)**: Reduces the dimensionality of data by transforming it into a set of linearly uncorrelated components

  - **t-Distributed Stochastic Neighbor Embedding (t-SNE)**: A non-linear dimensionality reduction technique, which is particularly good at visualizing high-dimensional data in two or three dimensions

Spark algorithms for feature extractors

Apache Spark, through **MLlib** (which stands for **Machine Learning Library**), offers a comprehensive tool for feature extraction, which is crucial for transforming raw data into a format that machine learning algorithms can effectively use. These tools are designed to handle large-scale datasets efficiently and cover a wide range of feature extraction needs. Here is an overview of the key feature extraction algorithms provided by Spark:

- **TF-IDF**:

  - **Purpose**: Reflects the importance of a term in the context of a document and the entire corpus

  - **Use case**: Used in text mining and information retrieval

- **Word2Vec**:

  - **Purpose**: Maps words to high-dimensional vectors

  - **Use case**: Useful for natural language processing tasks where words with similar meanings have similar representations

- **CountVectorizer**:

  - **Purpose**: Converts a collection of text documents to vectors of token counts

  - **Use case**: Often used in text classification and clustering for feature extraction from textual data

- **FeatureHasher**:

  - **Purpose**: Converts a set of categorical or numerical features into a feature vector of a specified dimension

  - **Use case**: Often used in classification and clustering for feature extraction from numerical or categorical data

In the next section, we will learn about feature transformers.

Code examples for feature extractors

Let's look at the different feature extractor algorithms and understand how they work through code examples.

TF-IDF

In the context of a table, a document corresponds to a single row within that table and the corpus refers to all the rows of the table.

The interpretations are as follows:

- The TF-IDF vector captures the importance of each term within the document relative to the entire corpus
- Terms that are unique to a specific document (rare across the corpus) receive higher TF-IDF scores

The following code snippet imports all the required functions, creates a Spark session, and creates a Spark DataFrame to implement TD-IDF:

```python
from pyspark.sql import SparkSession
from pyspark.ml.feature import HashingTF, IDF, Tokenizer
spark = SparkSession.builder.appName("TF-IDF Example").getOrCreate()

data = [
    (0, "This is the first document"),
    (1, "This document is the second document"),
    (2, "And this is the third one"),
    (3, "Is this the first document?"),
    (4, "The last document is the fifth one")
]

df = spark.createDataFrame(data, ["id", "text"])
df.show()
```

Here is the output:

```
+---+------------------------------------+
|id |text                                |
+---+------------------------------------+
|0  |This is the first document          |
|1  |This document is the second document|
|2  |And this is the third one           |
|3  |Is this the first document?         |
|4  |The last document is the fifth one  |
+---+------------------------------------+
```

Figure 3.1 – Data in a Spark DataFrame

We tokenize the text using `Tokenizer`:

```
tokenizer = Tokenizer(inputCol="text", outputCol="words")
words_df = tokenizer.transform(df)
```

We apply `HashingTF` to convert words into feature vectors:

```
hashingTF = HashingTF(inputCol="words",
    outputCol="rawFeatures", numFeatures=15)
featurized_df = hashingTF.transform(words_df)
```

We compute the IDF using `IDF`:

```
idf = IDF(inputCol="rawFeatures", outputCol="features")
idf_model = idf.fit(featurized_df)
tfidf_df = idf_model.transform(featurized_df)
```

Finally, we select the relevant columns, including the TF-IDF features.

```
result_df = tfidf_df.select("id", "text", "features")
result_df.show(truncate=False,vertical=False)
```

Here is the output:

```
+----+----+------------------------------------+----------------------------------------------------------------------------------------------------------------+
|    | id | text                               | features                                                                                                       |
+----+----+------------------------------------+----------------------------------------------------------------------------------------------------------------+
| 0  | 0  | This is the first document         | (15,[2,3,4,10,13],[0.0,0.1823215567939546,0.0,0.4054651081081644,0.6931471805599453])                          |
| 1  | 1  | This document is the second document| (15,[2,3,4,6,10],[0.0,0.1823215567939546,0.0,0.6931471805599453,0.8109302162163288])                           |
| 2  | 2  | And this is the third one          | (15,[0,1,2,3,4,6],[0.6931471805599453,0.6931471805599453,0.0,0.1823215567939546,0.0,0.6931471805599453])       |
| 3  | 3  | Is this the first document?        | (15,[1,2,3,4,13],[0.6931471805599453,0.0,0.1823215567939546,0.0,0.6931471805599453])                           |
| 4  | 4  | The last document is the fifth one | (15,[0,2,4,9,10],[0.6931471805599453,0.0,0.0,1.0986122886681098,0.4054651081081644])                           |
+----+----+------------------------------------+----------------------------------------------------------------------------------------------------------------+
```

Figure 3.2 – TF-IDF features in a Spark DataFrame

Word2Vec

The following code snippet imports all the required functions, creates a Spark session, and creates a Spark DataFrame for Word2Vec:

```
from pyspark.sql import SparkSession
from pyspark.ml.feature import Word2Vec, Tokenizer

# Create a Spark session
spark = SparkSession.builder.appName("Word2Vec Example").getOrCreate()

# Sample data (documents with tokenized words)
data = [
    (0, ["apple", "banana", "orange", "grape"]),
    (1, ["apple", "banana", "cherry", "pear"]),
```

```
    (2, ["banana", "cherry", "grape", "kiwi"]),
    (3, ["apple", "pear", "kiwi", "orange"]),
    (4, ["cherry", "grape", "kiwi", "orange"])
]

# Create a DataFrame from the sample data
df = spark.createDataFrame(data, ["id", "words"])
df.show()
```

Here is the output:

```
+---+----------------------------+
|id |words                       |
+---+----------------------------+
|0  |[apple, banana, orange, grape]|
|1  |[apple, banana, cherry, pear] |
|2  |[banana, cherry, grape, kiwi] |
|3  |[apple, pear, kiwi, orange]   |
|4  |[cherry, grape, kiwi, orange] |
+---+----------------------------+
```

Figure 3.3 – Data in a Spark DataFrame

The vectorSize parameter was set to 3, meaning that each word would be represented by a vector of three dimensions.

```
word2vec = Word2Vec(vectorSize=3, minCount=0,
    inputCol="words", outputCol="features")
```

The model learned semantic relationships between words based on their co-occurrence patterns in the dataset. The resulting features column contains Word2Vec embeddings for each document. These embeddings capture the context and meaning of words.

```
model = word2vec.fit(df)
result = model.transform(df)
result.select("id", "features").show(truncate=False)
```

```
+---+------------------------------------------------------------------+
|id |features                                                          |
+---+------------------------------------------------------------------+
|0  |[0.09411134896799922,-0.04479349683970213,-0.004068313166499138]  |
|1  |[0.05274438951164484,-0.010796030052006245,-0.009934199508279562] |
|2  |[0.02878595981746912,0.008933417033404112,0.01161392591893673]    |
|3  |[8.476478978991508E-5,-0.0047785197384655476,0.013960199896246195]|
|4  |[-0.007121491711586714,0.017444119323045015,0.036641841754317284] |
+---+------------------------------------------------------------------+
```

Figure 3.4 – WordVec features in Spark DataFrame

We can interpret that each row (document) now has a vector of Word2Vec features. These embeddings can be useful for various NLP tasks, such as finding synonyms, clustering related words, or even training downstream models.

CountVectorizer

The following code snippet imports all the required functions, creates a Spark session, and creates a Spark DataFrame for CountVectorizer:

```python
from pyspark.sql import SparkSession
from pyspark.ml.feature import CountVectorizer, Tokenizer
spark = SparkSession.builder.appName(
    "CountVectorizer Example").getOrCreate()
data = [
    (0, ["apple", "banana", "orange", "grape"]),
    (1, ["apple", "banana", "cherry", "pear"]),
    (2, ["banana", "cherry", "grape", "kiwi"]),
    (3, ["apple", "pear", "kiwi", "orange"]),
    (4, ["cherry", "grape", "kiwi", "orange"])
]
df = spark.createDataFrame(data, ["id", "words"])
df.show(truncate=False)
```

Here is the output:

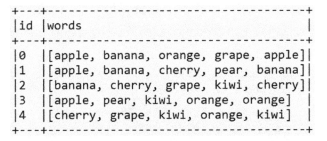

Figure 3.5 – Data in a Spark DataFrame

Here are the key code interpretations:

- The CountVectorizer model: We used the CountVectorizer model to generate word frequency counts for each document. The vocabSize parameter was set to 7 (0 to 6 – a total of 7 words), meaning that the vocabulary size (unique words) would be limited to 6. The model counted how many times each word appeared in each document.

- Word frequency counts (features): The resulting features column contains word frequency counts for each document. For example, if the word "apple" appeared twice in the first document, its count would be 2.

Each row (document) now has a vector of word counts:

```
cv = CountVectorizer(inputCol="words",
    outputCol="features", vocabSize=7)
model = cv.fit(df)
result = model.transform(df)
result.select("id", "features").show(truncate=False)
```

Here is the output:

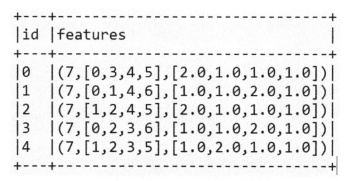

```
+---+------------------------------------+
|id |features                            |
+---+------------------------------------+
|0  |(7,[0,3,4,5],[2.0,1.0,1.0,1.0])|
|1  |(7,[0,1,4,6],[1.0,1.0,2.0,1.0])|
|2  |(7,[1,2,4,5],[2.0,1.0,1.0,1.0])|
|3  |(7,[0,2,3,6],[1.0,1.0,2.0,1.0])|
|4  |(7,[1,2,3,5],[1.0,2.0,1.0,1.0])|
+---+------------------------------------+
```

Figure 3.6 – Vectorized features in a Spark DataFrame

Let's take the first document as an example:

- Original text: ["apple", "banana", "orange", "grape", "apple"]
- Word counts: {"apple": 1, "banana": 1, "orange": 1, "grape": 1, "cherry": 0, "kiwi": 0}

The vector representation for this document might be [1, 1, 1, 1, 0, 0].

> **Note**
>
> Note that each token is assigned a random index. For example, the "apple" index is 0, whereas the "kiwi" index is 2.

CountVectorizer features are useful for various NLP tasks:

- **Text classification**: Representing documents as feature vectors for machine learning models
- **Clustering**: Grouping similar documents based on word frequencies
- **Topic modeling**: Identifying dominant topics within a corpus

Remember that CountVectorizer captures the occurrence frequency of words, which can be valuable for understanding document content and similarity

FeatureHasher

The following code snippet imports all the required functions, creates a Spark session, and creates a Spark DataFrame for FeatureHasher:

```
from pyspark.sql import SparkSession
from pyspark.ml.feature import FeatureHasher, Tokenizer
spark = SparkSession.builder.appName(
    "FeatureHasher Example").getOrCreate()
data = [
    (0, "apple", "banana", "orange"),
    (1, "apple", "banana", "cherry"),
    (2, "banana", "cherry", "grape"),
    (3, "apple", "pear", "kiwi"),
    (4, "cherry", "grape", "kiwi")
]
df = spark.createDataFrame(data,
    ["id", "feature1", "feature2", "feature3"])
df.show(truncate=False)
```

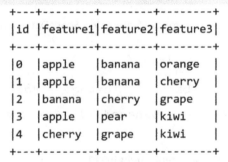

```
+---+--------+--------+--------+
|id |feature1|feature2|feature3|
+---+--------+--------+--------+
|0  |apple   |banana  |orange  |
|1  |apple   |banana  |cherry  |
|2  |banana  |cherry  |grape   |
|3  |apple   |pear    |kiwi    |
|4  |cherry  |grape   |kiwi    |
+---+--------+--------+--------+
```

Figure 3.7 – Data in a Spark DataFrame

Let's look at the various code components to understand them:

- **The FeatureHasher model:** We used the FeatureHasher model to generate hashed representations of the categorical features. The `numFeatures` parameter was set to `10`, meaning that each feature would be hashed into a vector of 10 dimensions. Hashing is a technique that maps categorical values to fixed-size vectors.

- **Hashed features:** The resulting features column contains hashed vectors for each document:

```
hasher = FeatureHasher(
    inputCols=["feature1", "feature2", "feature3"],
    outputCol="features", numFeatures=10)
result = hasher.transform(df)
result.select("id","features").show(truncate=False)
```

```
+---+--------------------------+
|id |features                  |
+---+--------------------------+
|0  |(10,[3,7,9],[1.0,1.0,1.0])|
|1  |(10,[3,6,7],[1.0,1.0,1.0])|
|2  |(10,[0,4,7],[1.0,1.0,1.0])|
|3  |(10,[0,7,8],[1.0,1.0,1.0])|
|4  |(10,[0,2,6],[1.0,1.0,1.0])|
+---+--------------------------+
```

Figure 3.8 – FeatureHasher in a Spark DataFrame

The interpretations are as follows:

- Each row (document) now has a vector of hashed features
- Hashing allows us to represent categorical features numerically
- Similar categorical values might map to similar vectors, but the exact mapping is determined by the hashing function

FeatureHasher is used in the following scenarios:

- It is useful when dealing with high-cardinality categorical features (many unique values)
- It's commonly used in machine learning pipelines to convert categorical data into a format suitable for algorithms

This concludes the section covering examples of feature extractors.

Working with feature transformers

In this section, we will understand what a feature transformer is, why it is very important in machine learning, and its various techniques. A feature transformer refers to the techniques or tools used to modify raw data into a more suitable format for analysis, particularly for use in training machine learning models. The transformation of features is a critical step in the data preprocessing phase, as it can significantly influence the performance of the models.

As discussed earlier, in Apache Spark's MLlib, **feature transformers** are essential tools for preprocessing data, transforming it into a format that is more suitable for use in machine learning algorithms. These transformers perform various operations on the dataset, such as scaling, normalization, conversion, and extraction, to make the data more compatible and effective for modeling.

The key aspects of feature transformers

The goal of feature transformers is to turn raw, potentially unwieldy data into a refined, structured format that can be effectively used by machine learning algorithms to make accurate predictions and uncover meaningful insights.

Here are some of the key aspects of feature transformers:

- **Normalization and standardization**: Adjust the scale of numerical features to a standard range or distribution. This is crucial for many machine learning algorithms that are sensitive to the scale of input data, such as support vector machines or k-nearest neighbors.

- **Encoding categorical variables**: Convert categorical data into numerical formats, as most machine learning models require numerical input. Techniques include one-hot encoding, label encoding, and binary encoding.

- **Handling missing values**: Deal with missing data in the dataset, which can affect model performance. Strategies to handle missing values include imputation (replacing missing values with mean, median, or another statistic), creating indicators for missing values, or omitting rows or columns with missing values.

- **Discretization**: Convert continuous features into categorical features through binning or bucketing, which can be beneficial for certain types of models or to handle non-linear relationships.

- **Feature creation and extraction**: Generate new features from existing ones to better capture underlying patterns in data, such as creating polynomial features or interaction terms.

- **Temporal and spatial feature engineering**: For time series or spatial data, extract relevant features that capture temporal or spatial dynamics (for example, lag features, moving averages, and spatial grids).

- **Dimensionality reduction**: Reducing the feature space can help improve the performance of machine learning models by reducing the risk of overfitting and decreasing the computational complexity.

Use cases and Spark algorithms for feature transformers

Feature transformers are pivotal in preparing data for machine learning models, and their applications span a wide range of industries and scenarios. Here are some practical use case examples where feature transformers are critically important:

- **Finance – credit scoring**:
  - **Normalization/standardization**: Financial variables such as income, loan amount, and age might be on vastly different scales. Using StandardScaler ensures that these features contribute equally to a model predicting creditworthiness.
  - **One-hot encoding**: Transforms categorical variables such as employment status or education level into a format suitable for machine learning models.

- **E-commerce – recommendation systems**:

  - **TF-IDF vectorization**: In a content-based recommendation system, product descriptions can be transformed into TF-IDF vectors to understand and compare textual content

  - **MinMaxScaler**: User ratings, often on a scale of 1 to 5, can be normalized to ensure consistency across different scales or platforms

- **Healthcare – patient risk stratification**:

  - **Imputation**: Handle missing values in patient records, which is common in medical datasets, by imputing missing values with the mean or median

  - **PCA (Principal Component Analysis)**: Reduce the dimensionality of patient data (such as various biomarkers) to identify key factors that influence health outcomes

- **Manufacturing – predictive maintenance**:

  - **Binning/bucketing**: Transform continuous sensor readings such as temperature or pressure into discrete categories to simplify the detection of abnormal conditions

  - **Polynomial features**: Create interaction terms between different sensor readings to capture complex relationships that might predict equipment failure

- **Real estate – property price prediction**:

  - **Feature creation**: Generate new features such as the age of a property or distance to the nearest metro station from existing date or location data

  - **StandardScaler**: Normalize features such as the square footage, number of bedrooms, or property tax rates for regression models predicting property prices

- **Online advertising – click-through rate prediction**:

  - **One-hot encoding**: Transform categorical data such as the user location, device type, and advert category into a binary format, suitable for logistic regression models predicting the likelihood of advert clicks

Spark algorithms for feature transformers

Apache Spark's MLlib provides a robust suite of feature transformers that are essential for data preprocessing in machine learning workflows. Here are some notable examples of the feature transformers available in Spark:

- **VectorAssembler**: Combines multiple columns of data (features) into a single vector column. This is useful for combining raw features and features generated by other transformers into a single feature vector, which is a common requirement for Spark's machine learning algorithms.

- **StringIndexer**: Converts a column of string labels into a column of label indices. It's often used to transform categorical text data into numerical indices that machine learning models can process.

- **OneHotEncoder**: Maps a column of category indices to a column of binary vectors. This is particularly useful after using StringIndexer to further transform categorical variables into a format that machine learning algorithms can utilize.

- **Tokenizer and RegexTokenizer**: Splits text into individual words or tokens. These are fundamental steps in text analysis and NLP tasks.

- **StopWordsRemover**: Removes common, uninformative words from a set of tokenized words. This helps to reduce the dimensionality of text data and focus on more meaningful words.

- **Bucketizer**: Transforms a column of continuous features into a column of feature buckets, based on user-specified splits. This is useful for converting continuous variables into categorical ones.

- **StandardScaler**: Standardizes features by removing the mean and scaling to unit variance. This is crucial for many machine learning models that assume that input features are on similar scales.

- **MinMaxScaler**: Transforms each feature to a given range, often [0, 1]. This is useful when you need to normalize the features to a specific range without distorting differences in the ranges of values.

- **Normalizer**: Transforms a dataset of vector rows, normalizing each vector to have a unit norm. This is used when you want to scale the vector observations to be of unit norm, which can be useful in clustering algorithms.

- **PCA**: Reduces the dimensionality of data by projecting it onto a smaller set of orthogonal features. PCA is used in exploratory data analysis and to make predictive models more efficient.

- **PolynomialExpansion**: Expands your feature vectors into a polynomial space, which is formulated by an n-degree combination of original dimensions. This can help to capture interactions between original features.

Code examples for feature transformers

Now, let's look at feature transformers through some code examples.

VectorAssembler

The following code snippet creates a sample Spark DataFrame with four columns, also called features:

```
from pyspark.ml.feature import VectorAssembler
data = [
    (1, 2, 3, 4),
    (5, 6, 7, 8),
    (9, 10, 11, 12)
```

```
]
columns = ["feature1", "feature2", "feature3", "feature4"]
df = spark.createDataFrame(data, columns)
df.show()
```

```
+--------+--------+--------+--------+
|feature1|feature2|feature3|feature4||
+--------+--------+--------+--------+
|       1|       2|       3|       4|
|       5|       6|       7|       8|
|       9|      10|      11|      12|
+--------+--------+--------+--------+
```

Figure 3.9 – Data in a Spark DataFrame

VectorAssembler is a feature transformer in PySpark that combines a given list of columns into a single vector column:

```
assembler = VectorAssembler(inputCols=columns, outputCol="features")
output_df = assembler.transform(df)
output_df.show()
```

```
+--------+--------+--------+--------+--------------------+
|feature1|feature2|feature3|feature4|            features|
+--------+--------+--------+--------+--------------------+
|       1|       2|       3|       4|   [1.0,2.0,3.0,4.0]|
|       5|       6|       7|       8|   [5.0,6.0,7.0,8.0]|
|       9|      10|      11|      12|[9.0,10.0,11.0,12.0]|
+--------+--------+--------+--------+--------------------+
```

Figure 3.10 – VectorAssembler features in a Spark DataFrame

The `features` column contains the combined values from all the columns.

The VectorAssembler features are used in the following use cases:

- **Feature engineering**: Combine raw features into a single vector for ML models
- **Logistic regression and decision trees**: Prepare input features for ML algorithms
- **Pipeline construction**: Assemble features in a pipeline
- **Sparse features**: Convert sparse features into dense vectors

StringIndexer

StringIndexer is a PySpark feature to convert categorical string columns in a DataFrame into numerical indices. This transformation is required because the majority of the machine learning algorithms cannot work directly with string data. StringIndexer assigns a unique index to each distinct string value in the input column and maps it to a new output column of integer indices.

The following code snippet creates a sample Spark DataFrame with two columns, with one of the columns as a categorical column:

```
from pyspark.ml.feature import StringIndexer
data = [ ("X", 100), ("X", 200), ("Y", 300), ("Y", 200),
         ("Y", 300), ("C", 400), ("Z", 100), ("Z", 100)]
columns = ["Categories", "Value"]
df = spark.createDataFrame(data, columns)
df.show()
```

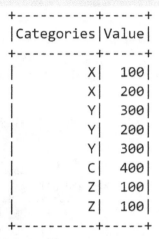

```
+----------+-----+
|Categories|Value|
+----------+-----+
|         X|  100|
|         X|  200|
|         Y|  300|
|         Y|  200|
|         Y|  300|
|         C|  400|
|         Z|  100|
|         Z|  100|
+----------+-----+
```

Figure 3.11 – Data in a Spark DataFrame

Let's see how StringIndexer works. StringIndexer calculates the input column's string values based on the frequency of the values in the dataset.

By default, the following applies:

- The most frequent label receives index 0

- The second most frequent label receives index 1, and so on

If two categories have the same frequency, the index value is assigned based on alphabetical order.

You can also use a custom ordering of the labels using the stringOrderType parameter:

```
indexer = StringIndexer(inputCol="Categories",
                        outputCol="Categories_Indexed")
indexerModel = indexer.fit(df)
indexed_df = indexerModel.transform(df)
indexed_df.show()
```

```
+----------+-----+------------------+
|Categories|Value|Categories_Indexed|
+----------+-----+------------------+
|        X|  100|               1.0|
|        X|  200|               1.0|
|        Y|  300|               0.0|
|        Y|  200|               0.0|
|        Y|  300|               0.0|
|        C|  400|               3.0|
|        Z|  100|               2.0|
|        Z|  100|               2.0|
+----------+-----+------------------+
```

Figure 3.12 – StringIndexer features in a Spark DataFrame

OneHotEncoder

OneHotEncoder is a feature transformer in PySpark that converts categorical variables (string labels) into a binary vector format. It creates a sparse vector where each unique category corresponds to a binary value (0 or 1).

The following code snippet creates a sample Spark DataFrame with two columns:

```
from pyspark.ml.feature import OneHotEncoder
data = [(0.0, 1.0), (1.0, 0.0), (2.0, 1.0)]
columns = ["input1", "input2"]
df = spark.createDataFrame(data, columns)
df.show()
```

```
+------+------+
|input1|input2|
+------+------+
|   0.0|   1.0|
|   1.0|   0.0|
|   2.0|   1.0|
+------+------+
```

Figure 3.13 – Data in a Spark DataFrame

The following code snippet processes the input columns by applying the OneHotEncoder model. It then transforms the input columns.

```
encoder = OneHotEncoder(
    inputCols=["input1", "input2"],
    outputCols=["output1", "output2"])
```

```
encoded_df = encoder.fit(df)
encoded_df = encoded_df.transform(df)
encoded_df.select("output1","output2").show(truncate=False)
```

```
+-------------+-------------+
|output1      |output2      |
+-------------+-------------+
|(2,[0],[1.0])|(1,[],[])    |
|(2,[1],[1.0])|(1,[0],[1.0])|
|(2,[],[])    |(1,[],[])    |
+-------------+-------------+
```

Figure 3.14 – OneHotEncoder features in a Spark DataFrame

Tokenizer and RegexTokenizer

The following code snippet creates a sample Spark DataFrame with two columns:

```
from pyspark.ml.feature import Tokenizer, RegexTokenizer
spark = SparkSession.builder.appName("TokenizerExample").getOrCreate()
data = [("Th+is is a sam+ple sent+ence.",)]
columns = ["text"]
df = spark.createDataFrame(data, columns)
df.show(truncate=False)
```

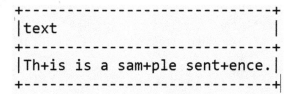

```
+-----------------------------+
|text                         |
+-----------------------------+
|Th+is is a sam+ple sent+ence.|
+-----------------------------+
```

Figure 3.15 – Data in a Spark DataFrame

Tokenizer splits the sentence into tokens, using whitespace as the delimiter.

On the other hand, RegexTokenizer splits the sentence into tokens using the custom " \W+ " pattern (which matches non-word characters):

```
tokenizer = Tokenizer(inputCol="text", outputCol="tokens")
tokenized_df = tokenizer.transform(df)
regex_tokenizer = RegexTokenizer(inputCol="text",
    outputCol="regex_tokens", pattern="\\+")
regex_tokenized_df = regex_tokenizer.transform(df)
tokenized_df.select("tokens").show(truncate=False)
regex_tokenized_df.select("regex_tokens").show(truncate=False)
```

```
+----------------------------------+
|tokens                            |
+----------------------------------+
|[th+is, is, a, sam+ple, sent+ence.]|
+----------------------------------+


+----------------------------------+
|regex_tokens                      |
+----------------------------------+
|[th, is is a sam, ple sent, ence.]|
+----------------------------------+
```

Figure 3.16 – Tokens in a Spark DataFrame

StopWordsRemover

The following code snippet creates a sample Spark DataFrame with two columns:

```python
from pyspark.sql import SparkSession
from pyspark.ml.feature import StopWordsRemover, Tokenizer
data = [("This is the first sentence.",),
        ("And here's another sentence.",),
        ("A third sentence for the DataFrame.",)]
columns = ["text"]
df = spark.createDataFrame(data, columns)
tokenizer = Tokenizer(inputCol="text", outputCol="words")
df = tokenizer.transform(df)
df.show(truncate=False)
```

```
+----------------------------------+------------------------------------------+
|text                              |words                                     |
+----------------------------------+------------------------------------------+
|This is the first sentence.       |[this, is, the, first, sentence.]         |
|And here's another sentence.      |[and, here's, another, sentence.]         |
|A third sentence for the DataFrame.|[a, third, sentence, for, the, dataframe.]|
+----------------------------------+------------------------------------------+
```

Figure 3.17 – Data in a Spark DataFrame

StopWordsRemover removes common stop words ("This," "is," "the," "And," "another", "A", and "for") from the input sentence.

The output contains only meaningful words ("first," "sentence," "another," "third", and "DataFrame").

The following code snippet applies the StopWordsRemover function to the input column to transform the filtering of stop words:

```python
stopwords_remover = StopWordsRemover(
    inputCol="words", outputCol="filtered_words")
```

```
filtered_df = stopwords_remover.transform(df)
filtered_df.select("filtered_words").show(truncate=False)
```

```
+---------------------------+
|filtered_words             |
+---------------------------+
|[first, sentence.]         |
|[another, sentence.]       |
|[third, sentence, dataframe.]|
+---------------------------+
```

Figure 3.18 – Filtered words in a Spark DataFrame

Stop words remover is used in text analysis and NLP to improve the quality of text data for NLP tasks.

Bucketizer

Bucketizer is a feature transformer in PySpark that discretizes continuous features into buckets (bins). It assigns each value to a specific bucket based on specified splits (boundaries). Bucketizing continuous features can improve model performance by capturing non-linear relationships.

The following code snippet creates a sample Spark DataFrame with two columns:

```
from pyspark.sql import SparkSession
from pyspark.ml.feature import Bucketizer
from pyspark.sql.functions import col
data = [(0, 1.5), (1, 2.5), (2, 3.5), (3, 4.5), (4, 5.5)]
columns = ["id", "value"]
df = spark.createDataFrame(data, columns)
df.show()
```

```
+---+-----+
| id|value|
+---+-----+
|  0|  1.5|
|  1|  2.5|
|  2|  3.5|
|  3|  4.5|
|  4|  5.5|
+---+-----+
```

Figure 3.19 – Data in a Spark DataFrame

The following snippet applies the bucketizer function to transform data:

```
splits = [0.0, 2.0, 4.0, float("inf")]
bucketizer = Bucketizer(splits=splits, inputCol="value",
    outputCol="bucket")
```

```
bucketized_df = bucketizer.transform(df)
bucketized_df.select("id", "value", "bucket").show()
```

```
+---+-----+------+
| id|value|bucket|
+---+-----+------+
|  0|  1.5|   0.0|
|  1|  2.5|   1.0|
|  2|  3.5|   1.0|
|  3|  4.5|   2.0|
|  4|  5.5|   2.0|
+---+-----+------+
```

Figure 3.20 – Bucketized values in a Spark DataFrame

The `value` column is bucketized into three buckets, based on the defined splits:

```
Bucket 0: (-∞, 2.0]
Bucket 1: (2.0, 4.0]
Bucket 2: (4.0, ∞)
```

StandardScaler

The StandardScaler is a feature transformer in PySpark that standardizes (scales) features by removing the mean and scaling to unit variance. It transforms continuous features to have zero mean and unit standard deviation. Standardization is essential for many machine learning algorithms that assume features are normally distributed.

The following code snippet creates a sample Spark DataFrame:

```
from pyspark.sql import SparkSession
from pyspark.ml.feature import StandardScaler
from pyspark.ml.linalg import Vectors
data = [(0, Vectors.dense([1.0, 0.1, -1.0]),),
        (1, Vectors.dense([2.0, 1.1, 1.0]),),
        (2, Vectors.dense([3.0, 10.1, 3.0]),)]
columns = ["id", "features"]
df = spark.createDataFrame(data, columns)
df.show()
```

```
+---+--------------+
| id|      features|
+---+--------------+
|  0|[1.0,0.1,-1.0]|
|  1| [2.0,1.1,1.0]|
|  2|[3.0,10.1,3.0]|
+---+--------------+
```

Figure 3.21 – Data in a Spark DataFrame

The following code snippet transforms the input columns into scaled features by applying the `StandardScaler` function:

```
scaler = StandardScaler(
    inputCol="features", outputCol="scaled_features",
    withStd=True, withMean=True)
scaler_model = scaler.fit(df)
scaled_df = scaler_model.transform(df)
scaled_df.select("id", "features",
    "scaled_features").show(truncate=False)
```

```
+---+--------------+-----------------------------------+
|id |features      |scaled_features                    |
+---+--------------+-----------------------------------+
|0  |[1.0,0.1,-1.0]|[-1.0,-0.6657502859356826,-1.0]|
|1  |[2.0,1.1,1.0] |[0.0,-0.4841820261350419,0.0]  |
|2  |[3.0,10.1,3.0]|[1.0,1.1499323120707245,1.0]   |
+---+--------------+-----------------------------------+
```

Figure 3.22 – Scaled features in a Spark DataFrame

The StandardScaler standardizes the `features` column by removing the mean and scaling to unit variance. The `scaled_features` column contains the standardized values.

MinMaxScaler

This linearly transforms each feature to fit within the specified minimum and maximum values. MinMax scaling is useful for algorithms that are sensitive to feature scaling, such as gradient-based optimization methods.

The following code snippet creates a sample Spark DataFrame:

```
from pyspark.sql import SparkSession
from pyspark.ml.feature import MinMaxScaler
from pyspark.ml.linalg import Vectors
data = [(0, Vectors.dense([1.0, 0.1, -1.0]),),
        (1, Vectors.dense([2.0, 1.1, 1.0]),),
        (2, Vectors.dense([3.0, 10.1, 3.0]),)]
columns = ["id", "features"]
df = spark.createDataFrame(data, columns)
df.show()
```

```
+---+--------------+
| id|      features|
+---+--------------+
|  0|[1.0,0.1,-1.0]|
|  1| [2.0,1.1,1.0]|
|  2|[3.0,10.1,3.0]|
+---+--------------+
```

Figure 3.23 – Data in a Spark DataFrame

The following code snippet transforms the input column by applying the `MinMaxScaler` function:

```
scaler = MinMaxScaler(
    inputCol="features", outputCol="scaled_features")
scaler_model = scaler.fit(df)
scaled_df = scaler_model.transform(df)
scaled_df.select("id", "features",
    "scaled_features").show(truncate=False)
```

```
+---+--------------+---------------+
|id |features      |scaled_features|
+---+--------------+---------------+
|0  |[1.0,0.1,-1.0]|(3,[],[])      |
|1  |[2.0,1.1,1.0] |[0.5,0.1,0.5]  |
|2  |[3.0,10.1,3.0]|[1.0,1.0,1.0]  |
+---+--------------+---------------+
```

Figure 3.24 – MinMax-scaled features in a Spark DataFrame

MinMaxScaler scales the `features` column to the `[0, 1]` range. The `scaled_features` column contains the scaled values.

Normalizer

It normalizes the feature vectors by dividing each value by the Euclidean norm (L2 norm). Normalization is useful when you want to compare vectors based on their direction rather than magnitude.

The following code snippet creates a sample Spark DataFrame:

```
from pyspark.sql import SparkSession
from pyspark.ml.feature import Normalizer
from pyspark.ml.linalg import Vectors
data = [(0, Vectors.dense([1.0, 0.1, -1.0]),),
        (1, Vectors.dense([2.0, 1.1, 1.0]),),
        (2, Vectors.dense([3.0, 10.1, 3.0]),)]
columns = ["id", "features"]
df = spark.createDataFrame(data, columns)
df.show()
```

```
+---+-------------+
| id|     features|
+---+-------------+
|  0|[1.0,0.1,-1.0]|
|  1| [2.0,1.1,1.0]|
|  2|[3.0,10.1,3.0]|
+---+-------------+
```

Figure 3.25 – Data in a Spark DataFrame

```
normalizer = Normalizer(inputCol="features",
    outputCol="normalized_features", p=2.0)
normalized_df = normalizer.transform(df)
normalized_df.select("id", "features",
    "normalized_features").show(truncate=False)
```

```
+---+-------------+-----------------------------------------------------------------------+
|id |features     |normalized_features                                                    |
+---+-------------+-----------------------------------------------------------------------+
|0  |[1.0,0.1,-1.0]|[0.7053456158585983,0.07053456158585983,-0.7053456158585983]|
|1  |[2.0,1.1,1.0] |[0.8025723539051279,0.4414147946478204,0.40128617695256397] |
|2  |[3.0,10.1,3.0]|[0.27384986857909926,0.9219612242163009,0.27384986857909926]|
+---+-------------+-----------------------------------------------------------------------+
```

Figure 3.26 – Normalized features in a Spark DataFrame

Normalizer scales each feature vector to have unit norm (an L2 norm). The `normalized_features` column contains the normalized vectors.

PCA

PCA is a dimensionality reduction technique used to transform high-dimensional data into a lower-dimensional space while preserving its essential information. It identifies the most important features (principal components) that explain the maximum variance in the data.

The following code snippet creates a sample Spark DataFrame:

```
from pyspark.sql import SparkSession
from pyspark.ml.feature import PCA
from pyspark.ml.linalg import Vectors
data = [(0, Vectors.dense([1.0, 0.1, -1.0])),
        (1, Vectors.dense([2.0, 1.1, 1.0])),
        (2, Vectors.dense([3.0, 10.1, 3.0]))]
columns = ["id", "features"]
df = spark.createDataFrame(data, columns)
df.show()
```

```
+---+-------------+
| id|     features|
+---+-------------+
|  0|[1.0,0.1,-1.0]|
|  1| [2.0,1.1,1.0]|
|  2|[3.0,10.1,3.0]|
+---+-------------+
```

Figure 3.27 – Data in a Spark DataFrame

The following code snippet applies the PCA function on the input column to fit and transform data into PCA features:

```
pca = PCA(k=2, inputCol="features", outputCol="pca_features")
pca_model = pca.fit(df)
pca_df = pca_model.transform(df)
pca_df.select("id", "pca_features").show(truncate=False)
```

```
+---+----------------------------------------+
|id |pca_features                            |
+---+----------------------------------------+
|0  |[0.06466700238304013,-0.4536718845187466]|
|1  |[-1.6616789696362084,1.2840650302335732] |
|2  |[-10.870750062210382,0.19181523649833343]|
+---+----------------------------------------+
```

Figure 3.28 – PCA features in a Spark DataFrame

The PCA model reduces the dimensionality of the features column to two principal components. The pca_features column contains the transformed data in the lower-dimensional space.

PolynomialExpansion

It allows us to capture non-linear relationships between features. PolynomialExpansion in PySpark creates new features by raising existing features to specified degrees.

The following code snippet creates a sample Spark DataFrame:

```
from pyspark.sql import SparkSession
from pyspark.ml.feature import PolynomialExpansion
from pyspark.ml.linalg import Vectors
data = [
    (0, Vectors.dense([1.0, 2.0])),
    (1, Vectors.dense([2.0, 3.0])),
    (2, Vectors.dense([3.0, 4.0]))]
columns = ["id", "features"]
df = spark.createDataFrame(data, columns)
df.show()
```

```
+---+---------+
| id| features|
+---+---------+
|  0|[1.0,2.0]|
|  1|[2.0,3.0]|
|  2|[3.0,4.0]|
+---+---------+
```

Figure 3.29 – Data in a Spark DataFrame

The following code snippet applies the `PolynomialExpansion` function to transform the input column into polynomial features:

```
poly_expansion = PolynomialExpansion(inputCol="features",
    outputCol="expanded_features", degree=2)
expanded_df = poly_expansion.transform(df)
expanded_df.select("id", "expanded_features").show(truncate=False)
```

```
+---+------------------------+
|id |expanded_features       |
+---+------------------------+
|0  |[1.0,1.0,2.0,2.0,4.0]   |
|1  |[2.0,4.0,3.0,6.0,9.0]   |
|2  |[3.0,9.0,4.0,12.0,16.0]|
+---+------------------------+
```

Figure 3.30 – Polynomial features in a Spark DataFrame

The PolynomialExpansion generates new features by creating polynomial combinations of the original features. For example, the first row has [1.0, 2.0] features, and the expanded features include [1.0, 2.0, 1.0, 4.0, 0.0, 0.0].

In the next section, we will explore more about feature selectors.

Exploring feature selectors

In this section, we will see what a feature selector is, why it is very important in machine learning, and its various techniques. A feature selector in machine learning is a tool or method used to automatically select a subset of relevant features (variables and predictors) for use in model construction. This process is an essential part of the feature engineering phase and significantly impacts the performance of machine learning models.

Apache Spark's MLlib provides several feature selection tools that are crucial in selecting the most relevant features for model training. These feature selectors help enhance model performance, reduce complexity, and improve computational efficiency.

The key aspects of feature selectors

Feature selectors are typically used in the data preprocessing stage. After initial data cleaning and transformation, feature selectors are applied to choose the most relevant features before data is fed into a machine learning algorithm.

The selection process is often guided by statistical tests (such as the chi-squared test) or based on domain knowledge and data exploration.

Use cases and Spark algorithms for feature selectors

Here are the key feature selectors available in Apache Spark:

- **Chi-Squared Selector (ChiSqSelector):**

  - **Function:** This selects categorical features to use to predict a categorical label, based on the chi-squared statistical test. The test is used to determine the independence of two events, making it useful for selecting features that have the strongest relationship with the label.

  - **Use case:** Ideal for classification tasks where features are categorical. It helps to select those features that show a strong association with the label.

- **Vector slicer:**

  - **Function:** A feature transformer that selects a set of indices from a feature vector. It's useful for extracting features from a vector column.

  - **Use case:** Handy when you need to slice and extract features from a large feature vector, based on their indices. It can be used for manual feature selection based on domain knowledge or previous analysis.

- **RFormula:**

  - **Function:** Provides R-like formulae to define the model specification in terms of features and interaction terms. While not a feature selector in the traditional sense, it allows you to specify which features to include in a model.

  - **Use case:** Useful for creating complex models that include interactions between variables and applying transformations to variables.

- **UnivariateFeatureSelector**:

  - **Function**: UnivariateFeatureSelector is a feature selection technique used in machine learning to select the most relevant features from a dataset, based on their individual statistical properties and relationships with the target variable. It is a univariate feature selection method because it considers each feature independently without considering the interactions between features.

  - **Use case**: In classification and regression tasks, you can use the univariate feature selector to select the most informative features to improve model accuracy and reduce overfitting.

In tasks, it helps to identify relevant predictors and improve a model's predictive performance.

Code examples of feature selectors

Let's now look at the feature selector algorithm and walk through some of the code examples.

Chi-squared Selector

ChiSqSelector is a feature selector in PySpark that uses the chi-squared statistical test to select the most important features. It is commonly used for feature selection in classification problems with categorical features. The chi-squared test measures the independence between categorical variables.

The following code snippet imports the required libraries, creates a sample Spark DataFrame, and displays the output:

```
from pyspark.sql import SparkSession
from pyspark.ml.feature import ChiSqSelector
from pyspark.ml.linalg import Vectors
from pyspark.sql.functions import col
data = [(0, Vectors.dense([1.0, 0.1, -1.0]), 1.0),
        (1, Vectors.dense([2.0, 1.1, 1.0]), 0.0),
        (2, Vectors.dense([3.0, 10.1, 3.0]), 0.0)]
columns = ["id", "features", "label"]
df = spark.createDataFrame(data, columns)
df.show()
```

```
+---+---------------+-----+
| id|       features|label|
+---+---------------+-----+
|  0|[1.0,0.1,-1.0]|  1.0|
|  1| [2.0,1.1,1.0]|  0.0|
|  2|[3.0,10.1,3.0]|  0.0|
+---+---------------+-----+
```

Figure 3.31 – Data in a Spark DataFrame

The following code snippet fits the data using the `ChiSqSelector` algorithm to generate a model. This model is then used to transform the data to extract the features.

```
selector = ChiSqSelector(
    numTopFeatures=1, featuresCol="features",
    outputCol="selected_features", labelCol="label")
selector_model = selector.fit(df)
selected_df = selector_model.transform(df)
selected_df.select("id", "selected_features").show(truncate=False)
```

```
+---+-----------------+
|id |selected_features|
+---+-----------------+
|0  |[1.0]            |
|1  |[2.0]            |
|2  |[3.0]            |
+---+-----------------+
```

Figure 3.32 – ChiSqSelector-selected features in a Spark DataFrame

The ChiSqSelector selects the top k features based on the chi-squared test. In this example, we selected the single most important feature.

VectorSlicer

VectorSlicer is a feature transformer in PySpark that extracts a subset of features from a vector column. It allows you to select specific features based on their indices or names. The VectorSlicer is useful for creating new feature vectors with a subset of the original features.

The following code snippet imports the required libraries, creates a sample Spark DataFrame, and displays the output:

```
from pyspark.sql import SparkSession
from pyspark.ml.feature import VectorSlicer
from pyspark.ml.linalg import Vectors
data = [(0, Vectors.dense([1.0, 2.0, 3.0, 4.0, 5.0])),
        (1, Vectors.dense([2.0, 3.0, 4.0, 5.0, 6.0])),
        (2, Vectors.dense([3.0, 4.0, 5.0, 6.0, 7.0]))]
columns = ["id", "features"]
df = spark.createDataFrame(data, columns)
df.show()
```

```
+---+--------------------+
| id|            features|
+---+--------------------+
|  0|[1.0,2.0,3.0,4.0,...|
|  1|[2.0,3.0,4.0,5.0,...|
|  2|[3.0,4.0,5.0,6.0,...|
+---+--------------------+
```

Figure 3.33 – Data in a Spark DataFrame

The following code snippet transforms data using VectorSlicer to extract the features:

```
slicer = VectorSlicer(inputCol="features",
    outputCol="selected_features", indices=[1, 3, 4])
sliced_df = slicer.transform(df)
sliced_df.select("id", "selected_features").show(truncate=False)
```

```
+---+-----------------+
|id |selected_features|
+---+-----------------+
|0  |[2.0,4.0,5.0]    |
|1  |[3.0,5.0,6.0]    |
|2  |[4.0,6.0,7.0]    |
+---+-----------------+
```

Figure 3.34 – Vectorslicer features in a Spark DataFrame

The VectorSlicer selects specific features (indices 1, 3, and 4) from the original feature vector. The `selected_features` column contains the extracted features.

RFormula

RFormula is a feature transformer in PySpark that allows you to express transformations required to fit a dataset against an R model formula. It supports a limited subset of R operators, including ~, ., :, +, -, *, and ^. The RFormula is useful for creating features from a combination of existing columns.

The following code snippet imports the required libraries, creates a sample Spark DataFrame, and displays the output:

```
from pyspark.sql import SparkSession
from pyspark.ml.feature import RFormula
spark = SparkSession.builder.appName("RFormulaExample").getOrCreate()
data = [(1.0, 1.0, "a"),
        (0.0, 2.0, "b"),
        (0.0, 0.0, "a")]
```

```
columns = ["y", "x", "s"]
df = spark.createDataFrame(data, columns)
df.show()
```

```
+---+---+---+
|  y|  x|  s|
+---+---+---+
|1.0|1.0|  a|
|0.0|2.0|  b|
|0.0|0.0|  a|
+---+---+---+
```

Figure 3.35 – Data in a Spark DataFrame

The following code snippet transforms data using RFormula to extract the features:

```
rf = RFormula(formula="y ~ x + s")
model = rf.fit(df)
transformed_df = model.transform(df)
transformed_df.select("y", "x", "s", "features",
    "label").show(truncate=False)
```

```
+---+---+---+---------+-----+
|y  |x  |s  |features |label|
+---+---+---+---------+-----+
|1.0|1.0|a  |[1.0,1.0]|1.0  |
|0.0|2.0|b  |[2.0,0.0]|0.0  |
|0.0|0.0|a  |[0.0,1.0]|0.0  |
+---+---+---+---------+-----+
```

Figure 3.36 – RFormula features in a Spark DataFrame

The RFormula creates a feature vector based on the specified formula (y ~ x + s). The features column contains the combined features, and the label column contains the target variable.

UnivariateFeatureSelector

The UnivariateFeatureSelector is a feature selector in PySpark that selects the most important features based on univariate statistical tests. It evaluates each feature independently and ranks them according to their relevance to the target variable. The UnivariateFeatureSelector is useful for feature selection in classification or regression tasks.

The following code snippet imports the required libraries, creates a sample Spark DataFrame, and displays the output:

```
from pyspark.sql import SparkSession
from pyspark.ml.feature import UnivariateFeatureSelector
from pyspark.ml.linalg import Vectors
data = [(1.0, Vectors.dense([1.0, 0.1, -1.0])),
        (0.0, Vectors.dense([2.0, 1.1, 1.0])),
        (0.0, Vectors.dense([3.0, 10.1, 3.0]))]
columns = ["label", "features"]
df = spark.createDataFrame(data, columns)
df.show()
```

```
+-----+--------------+
|label|      features|
+-----+--------------+
|  1.0|[1.0,0.1,-1.0]|
|  0.0| [2.0,1.1,1.0]|
|  0.0|[3.0,10.1,3.0]|
+-----+--------------+
```

Figure 3.37 – Data in a Spark DataFrame

The following code snippet transforms data using UnivariateFeatureSelector to extract the features:

```
selector = UnivariateFeatureSelector(
    featuresCol="features", outputCol="selected_features")
selector.setFeatureType("continuous") \
    .setLabelType("categorical") \
    .setSelectionThreshold(1)
selected_df = selector.fit(df).transform(df)
selected_df.select(
    "label", "selected_features").show(truncate=False)
```

```
+-----+-----------------+
|label|selected_features|
+-----+-----------------+
|1.0  |[1.0]            |
|0.0  |[2.0]            |
|0.0  |[3.0]            |
+-----+-----------------+
```

Figure 3.38 – Univariate features in a Spark DataFrame

The UnivariateFeatureSelector selects the single most important feature based on statistical tests. In this example, we selected the feature with the highest relevance to the target variable.

Summary

Effective feature engineering significantly impacts model accuracy and efficiency. In this chapter, we looked at several of PySpark's feature transformation algorithms, along with their use cases and code examples. Similarly, we also learned about feature selection algorithms, their use cases, and code examples. Finally, we learned about feature extraction algorithms and their code examples. Data scientists can create informative features and build robust machine learning models. Remember to choose the appropriate techniques based on the problem domain, data characteristics, and model requirements.

In the next chapter, we will look into regression, a type of supervised learning.

Part 2:
Supervised Learning

In this part, you will explore the key concepts, algorithms, and applications of supervised learning – one of the most fundamental techniques in machine learning. Supervised learning involves training a model on labeled data, enabling it to make predictions or decisions based on new, unseen data. This part will provide you with a comprehensive understanding of how to build, evaluate, and optimize supervised learning models.

The chapters in this part will guide you through the essential methods and best practices for applying supervised learning to real-life problems. You will learn how to build regression and classification models, select appropriate algorithms, and improve model performance through feature engineering and model tuning.

This part contains the following chapters:

- *Chapter 4, Building a Regression System*
- *Chapter 5, Building a Classification System*

Building a Regression System

Supervised learning, at its core, involves training a model on a labeled dataset, where each example is paired with the correct output, enabling the model to learn the mapping from inputs to outputs. Linear regression is a statistical method that finds the linear equation that fits best to predict the dependent variable based on the values of the independent variables. Learning regression is crucial in the field of statistics and data science for several compelling reasons as it serves as a foundational tool for understanding relationships between variables, making predictions, and informing decision-making processes across various domains.

First, we'll gain general insights into regression and its benefits. Then, the focus will shift to a deep understanding of various regression algorithms. Each algorithm will be accompanied by real-world case studies and code examples illuminating their practical applications. We will learn how to formulate a regression problem, choose an appropriate model and algorithm, train and evaluate the model, and apply the model to new data. Several machine learning use cases can be solved using regression techniques. Many complex machine learning solutions contain regression as one of the components to solve one part of the problem.

We will cover the following topics in this chapter:

- Learning about regression
- Learning regression algorithms
- Evaluating the model's performance
- Improving the model's performance

By the end of this chapter, you will be able to understand and build practical insights into building a regression model using several algorithms and evaluate the model's output.

Technical requirements

You can find the code files for this chapter on GitHub at `https://github.com/PacktPublishing/Apache-Spark-for-Machine-Learning/tree/main/Chapter04`.

Learning about regression

In this section, we'll gain a general overview of regression, as well as review its benefits and practical use cases. This will help us have a solid foundation regarding regression and be able to identify regression use cases.

Regression overview

Regression is a type of supervised machine learning technique that allows you to predict a continuous or numerical output based on one or more input features. Regression models learn the relationship between the input and output variables from labeled data and use this relationship to make predictions for new or unseen data. Regression is related to machine learning because it is one of the fundamental methods for solving various prediction, estimation, and optimization problems. Regression can also help you understand how different features affect the output and how to improve the performance and accuracy of your model. You can use regression when you want to do the following:

- Explore how different variables are related to each other and how they affect a certain outcome
- Test hypotheses about the effects of one or more variables on another variable
- Control confounding factors that may influence the relationship between variables
- Forecast future trends or values based on historical data and current conditions

If you're thinking about solving the following problems, then you should think about regression techniques:

- Predicting the price of a house based on its size, location, and amenities
- Estimating the demand for a product or service based on historical data and market trends
- Forecasting the sales, revenue, or profit of a business based on various factors

Some use cases of regression in machine learning

The following are use cases of regression in several domains:

- **Business**: Regression can help you analyze customer behavior, optimize marketing strategies, evaluate policies, forecast sales, and detect anomalies
- **Medicine**: Regression can help you measure the effect of drugs, test hypotheses, adjust for confounding factors, and predict outcomes
- **Agriculture**: Regression can help you study the effect of fertilizer, water, and other factors on crop yields, as well as optimize the use of resources
- **Engineering**: Regression can help you test and improve the performance of machines, processes, and systems, and estimate the time to failure

- **Education**: Regression can help you assess the impact of socio-economic status, race, and other factors on educational achievement, as well as design effective interventions

- **Science**: Regression can help you explore the relationship between physical, chemical, biological, and other phenomena, as well as test theories and models

Pitfalls of regression techniques

Regression analysis is a powerful statistical method for estimating the relationships among variables. However, it has several potential pitfalls and assumptions that must be considered to ensure valid and reliable results. Here are some common pitfalls of regression analysis:

- **Violation of assumptions**: Regression analysis, especially linear regression, is based on several key assumptions, including linearity, independence of errors, homoscedasticity (constant variance of error terms), and normal distribution of error terms. Violating these assumptions can lead to biased or misleading results.

- **Multicollinearity**: This occurs when independent variables in the regression model are highly correlated with each other. Multicollinearity can make it difficult to assess the effect of individual variables on the dependent variable and can inflate the variance of the coefficient estimates, making them unstable and unreliable.

- **Overfitting**: Overfitting happens when the model is too complex and captures the noise in the data along with the underlying pattern. This makes the model perform well on the training data but poorly on new, unseen data. It's important to strike a balance between model complexity and generalization.

- **Underfitting**: Conversely, underfitting occurs when the model is too simple to capture the underlying structure of the data, leading to poor performance both on the training data and on new data.

- **Extrapolation**: Making predictions outside the range of the data used to fit the model (extrapolation) can be highly unreliable as the relationship between variables may not hold outside the observed data range.

- **Measurement error**: Errors in measuring the independent or dependent variables can lead to biased and inconsistent estimates.

- **Endogeneity**: This occurs when an independent variable is correlated with the error term. It often arises due to omitted variable bias, measurement error, or simultaneity (where cause and effect between variables are not distinguishable). Endogeneity can lead to biased and inconsistent parameter estimates.

- **Sample size**: Too small a sample size can lead to a lack of statistical power to detect significant relationships, while too large a sample size can detect trivial effects as significant.

- **Non-linear relationships**: Linear regression models can fail to capture non-linear relationships unless they are properly transformed or non-linear models are used.

- **Data quality**: Poor data quality, including outliers, missing values, and incorrect data, can lead to misleading regression results.

To mitigate these pitfalls, it's crucial to conduct thorough exploratory data analysis, test the assumptions of your regression model, consider using regularization techniques to prevent overfitting, and explore alternative modeling approaches when necessary.

Next, we will learn more about regression algorithms.

Learning regression algorithms

Now, let's understand various regression algorithms. The following regression algorithms are available in Apache Spark:

- Linear regression
- Generalized linear regression (note that linear regression and generalized linear regression are two different algorithms)
- Decision tree regression
- Random forest regression
- Gradient-boosted tree regression
- Survival regression
- Factorization machine regressor

We will review them in detail next.

Linear regression

Linear regression is a fundamental algorithm in statistics and machine learning that's used to model the relationship between a dependent variable and one or more independent variables by fitting a linear equation to observed data. The main outcome of linear regression is to find the best-fitting straight line through the observed data points.

Linear regression can be analogously described as a chef creating a recipe for a signature dish. Imagine a chef trying to perfect a recipe for a cake. The final taste of the cake (the outcome or dependent variable) is influenced by various ingredients (the independent variables or predictors), such as flour, sugar, eggs, and butter.

In this analogy, the chef experiments with different quantities of these ingredients to find the perfect balance that yields the best-tasting cake. Each ingredient has a different impact on its taste, much like how each predictor variable in linear regression has a different weight or coefficient.

Just as the chef knows that adding too much flour might make the cake too dense (a form of bias) or that the perfect level of sweetness can vary slightly each time (capturing the random error), linear regression accounts for these variations with the regression coefficients and the error term, aiming to minimize the difference between the predicted and actual outcomes.

In summary, just as the chef iterates on the recipe to find the right mix that will consistently produce a delicious cake, linear regression iterates over the data to find the right coefficients that best predict the outcome:

Figure 4.1 – Analogy of linear regression

Applications

There are three notable applications:

- **Finance and economics**:
  - **Risk assessment**: It predicts loan defaults and evaluates the risk factors affecting investment portfolios
  - **Demand forecasting**: Used to predict consumer demand for products and services
- **Real estate**:
  - **Property pricing**: Extensively used to predict property prices based on features such as location, size, number of bedrooms, and so on

- **Marketing**:

  - **Sales forecasting**: Used to predict future sales based on various predictors, such as advertising spend, market conditions, and competitor actions

  - **ROI on advertising**: Linear regression helps in measuring the **return on investment** (**ROI**) of different marketing channels by correlating marketing spend with sales increases, enabling marketers to allocate budgets more effectively

Benefits

The top three benefits of linear regression, highlighting its widespread use and significance in data analysis and predictive modeling, are as follows:

- **Simplicity and interpretability**: Linear regression models are easy to understand, making them highly interpretable

- **Efficiency**: Due to its computational simplicity, linear regression is efficient, especially for large datasets

- **Foundation base model for understanding more complex models**: Linear regression serves as a baseline model for many advanced modeling techniques and algorithms

Limitations

Among the limitations of linear regression, the top three that often pose significant challenges are as follows:

- **Linearity assumption**: Linear regression assumes a linear relationship between the independent and dependent variables. This is a critical limitation because many real-world phenomena exhibit non-linear relationships.

- **Influence of outliers**: The presence of outliers can significantly impact the performance of a linear regression model as they can skew the regression line, leading to a poor fit.

- **Multicollinearity**: Multicollinearity occurs when two or more independent variables in a multiple linear regression model are highly correlated, making it difficult to isolate the individual effect of each independent variable from the dependent variable. This can lead to instability. Estimates of the coefficients, where small changes in the data can lead to large changes in the model, make interpretations and predictions less reliable.

Generalized linear regression

The generalized linear regression method uses the **generalized linear model** (**GLM**) algorithm, which extends traditional linear regression to accommodate response variables that do not follow a normal distribution. GLM consists of three components: a random component specifying the distribution of the response variable, a systematic component defining the linear predictor of the independent variables, and a link function connecting the expected value of the response to the linear predictor. This framework allows various data types (for example, binary, count, and so on) to be modeled using distributions such as binomial or Poisson, making GLM versatile for diverse analytical needs across fields such as biology, economics, and social sciences.

Imagine you're a chef trying to perfect a soup recipe. In a basic soup recipe (linear regression), you might adjust simple ingredients such as salt and pepper, which directly influence the taste in a straightforward manner: adding a teaspoon of salt makes the soup predictably saltier, and this relationship is linear and direct.

Now, consider that you're experimenting with a more complex recipe that involves ingredients that don't have a straightforward impact on the soup's taste, such as spices that need to be balanced with each other, or ingredients that change flavor under different cooking conditions (generalized linear regression). In this scenario, the effect of adding a spice might depend on the presence and quantity of other spices, and the cooking temperature might change the flavors in non-linear ways.

In generalized linear regression, you're still trying to perfect your soup (predict an outcome), but now, you're working with a wider range of ingredients (predictors) that might have complex, non-linear relationships with the taste (outcome). Just as you'd adjust your cooking technique for different ingredients (such as simmering tomatoes to sweeten them or roasting spices to enhance their flavors), generalized linear regression adjusts for the different relationships between predictors and the outcome by using link functions and different types of distribution (such as binomial for yes/no outcomes, Poisson for count data, and so on), allowing for a more flexible and nuanced approach to "cooking" the perfect predictive model:

Figure 4.2 – Analogy of GLM

Applications

GLMs are widely used across different fields due to their flexibility in handling various types of data. Their top three applications are as follows:

- **Economics**: In economics, GLMs can model consumer behavior, labor market dynamics, and other phenomena where the relationships between variables are not necessarily linear or normally distributed.

- **Insurance**: GLMs are applied to model claim frequencies and amounts, often following distributions such as Poisson or Gamma rather than the normal distribution. These models help set premiums, understand risk factors, and manage reserves.

- **Environmental science**: In environmental science, GLMs are used to model count data or rates, such as the number of species in an area (Poisson regression) or the time until a certain event occurs (exponential regression). These applications are crucial for understanding biodiversity, predicting natural events, and managing resources.

Benefits

GLMs extend the utility and applicability of linear regression by allowing for more flexibility in data modeling. Let's consider the top three benefits of GLMs:

- **Flexibility in modeling different types of data**: GLMs can accommodate various types of response variables, including binary, count, and continuous outcomes, by allowing for different distributions (for example, binomial for binary data, Poisson for count data, and so on).

- **Ability to handle non-normal error distributions**: Unlike traditional linear regression, which assumes that the residuals are normally distributed, GLMs can model data with error distributions that are not normal, such as binomial, Poisson, and gamma distributions. This capability allows for more accurate modeling of the underlying data structure, leading to better predictions and inferences.

- **Direct modeling of the response variable's mean**: Through link functions, GLMs enable the direct modeling of the response variable's mean as a function of the predictor variables. This direct relationship allows us to interpret the effects of predictors on the response variable.

Limitations

GLMs are powerful and flexible tools for statistical modeling, but they come with certain limitations that are important to consider:

- **Assumption regarding the correct distribution and link function**: A key challenge in using GLMs is selecting the appropriate distribution for the response variable and the correct link function that relates the linear predictor to the mean of the distribution.

- **Sensitivity to outliers and influential data points**: Like linear regression, GLMs can be sensitive to outliers and influential observations.

- **Complexity in interpretation and communication**: While GLMs extend the flexibility of linear regression, they can also introduce additional complexity, especially with non-canonical link functions or when modeling non-standard distributions. This complexity can make interpreting model coefficients and predictions more challenging.

Decision tree regression

Decision tree regression involves segmenting the predictor space into distinct and non-overlapping regions. For a given predictor value, the mean or median output of the training observations in the region to which the value belongs is returned as the prediction. This method employs a tree-like model of decisions and their possible consequences, including chance event outcomes and resource costs.

Imagine planning a road trip across the country with several stops, and you're trying to estimate the total travel time. You start by considering a simple decision: if the distance to the next stop is less than 100 miles, you predict the travel time will be short. If it's more than 100 miles, you predict it will be long. This initial decision is akin to the first "split" in a decision tree, where you divide your route based on a key factor influencing your total travel time.

As you plan further, you realize there are more factors to consider at each stop, such as traffic, road conditions, and weather. For instance, if the distance is less than 100 miles but heavy traffic is expected, you adjust your travel time prediction upwards. Conversely, for long distances with clear weather and good road conditions, you might predict a relatively swift journey. Each of these considerations represents further "splits" in your decision tree, where at every junction (or node), based on the conditions (or features), you define to refine your travel time estimate.

In a decision tree regression, just like in your road trip planning scenario, the model makes a series of binary decisions or splits based on the features of the data. Each split aims to group data points (or segments of your trip) with similar outcomes (travel times), refining the predictions at each step. The final "leaves" of the tree represent the segments of your journey for which you've accounted for all relevant conditions, and you have a specific travel time prediction based on the combination of factors leading to that leaf.

This decision-making process, much like planning the detailed segments of your road trip while considering various conditions, illustrates the essence of decision tree regression: breaking down a complex prediction problem into a series of simpler, conditional decisions to arrive at a more accurate and nuanced estimate.

Applications

The decision tree algorithm has various applications:

- **Real estate pricing**: Predicting property prices based on features such as location, size, and amenities

- **Energy consumption**: Forecasting energy usage in buildings from historical usage data and environmental factors
- **Supply chain optimization**: Predicting demand for products to optimize inventory levels

Benefits

The benefits are as follows:

- **Interpretability**: Easy to understand and visualize, even for non-technical users
- **Non-linearity**: Can capture non-linear relationships without needing transformations
- **No distribution assumption**: Does not assume any specific distribution of the data

Limitations

Apart from the benefits, there are also some limitations:

- **Overfitting**: Prone to overfitting, especially with complex trees
- **Instability**: Small changes in data can lead to vastly different trees
- **Bias**: Decision trees can be biased toward dominant classes, affecting regression accuracy

Random forest regression

Random forest regression is an ensemble learning method that works by creating multiple decision trees during training and then producing the average or median prediction of the individual trees. It merges the simplicity of decision trees with flexibility, enhancing accuracy and robustness.

Imagine you're planning a large outdoor event and you're trying to predict the number of attendees. Relying on just your own or a single friend's guess might be too simplistic and prone to error. So, you decide to form a committee where each member brings their unique perspective based on different factors such as weather forecasts, day of the week, the popularity of the event, past event attendance, and promotional efforts.

Each committee member uses their specific set of considerations to make a prediction. One member might focus on the weather and day of the week, predicting higher attendance on a sunny weekend. Another might consider the event's theme and its historical popularity, making a different prediction. This process is akin to how a decision tree in a random forest regression operates, focusing on subsets of features to make a prediction.

After each committee member has made their prediction, you don't just take any single prediction as the final estimate. Instead, you aggregate all the predictions, possibly taking the average, to come up with a more robust, collective forecast of the attendance. This method is likely to be more accurate and reliable because it combines diverse perspectives and reduces the influence of any outlier predictions.

This committee-based approach mirrors the random forest regression algorithm. In this case, many decision trees (the committee members) are grown, each considering random subsets of features and data points to predict the outcome (event attendance). The final prediction is an aggregation (that is, average) of all these individual tree predictions, leading to a more accurate and stable model that's less prone to the overfitting that might occur if you were to rely on a single decision tree (or committee member's guess). This ensemble method leverages the power of the "wisdom of the crowd" to arrive at a more reliable prediction, much like your committee collectively forecasts event attendance more effectively than any single member could.

Applications

Random forest regression has the following applications:

- **Market analysis and prediction**: Estimating future sales, stock prices, and market trends based on historical data
- **Medical diagnostics**: Predicting disease progression and patient outcomes based on clinical variables
- **Environmental modeling**: Forecasting air quality, temperature, and other climate variables using historical sensor data

Benefits

Let's review the benefits:

- **Accuracy**: Combining multiple trees reduces variance and improves prediction accuracy
- **Robustness**: Less prone to overfitting than individual decision trees
- **Feature importance**: Provides insights into which predictors are most influential in determining the outcome

Limitations

The limitations associated with random forest regression are as follows:

- **Complexity**: More computationally intensive than single decision trees, requiring more resources
- **Interpretability**: Its ensemble nature makes the model more challenging to interpret than a single decision tree
- **Tuning required**: Requires parameters (such as the number of trees) to be tuned carefully to avoid overfitting and ensure optimal performance

Gradient-boosted tree regression

Gradient-boosted tree regression is an ensemble technique that builds multiple decision trees sequentially, where each tree attempts to correct the errors of the previous ones. By combining weak learners, typically shallow trees, it gradually improves accuracy through iterative optimization, minimizing a loss function.

Imagine you're coaching a team of novice archers to hit a distant target. Your first archer takes a shot, but the arrow falls short. You note how far and in which direction it missed, offering this insight to the second archer. The second archer adjusts their aim based on the first's mistake but still misses, albeit by a smaller margin. Each subsequent archer learns from the cumulative errors of all previous archers, adjusting their aim to correct those errors. Over time, the shots get closer and closer to the target as each archer builds on the collective knowledge of their predecessors' attempts.

In this analogy, each archer represents an individual decision tree in a gradient-boosted tree regression model. The first tree (archer) makes a rough estimate (shot), aiming to predict the target value (hit the target). The prediction error (distance and direction of the miss) is then used to inform the next tree (next archer's shot). Each new tree focuses on correcting the residual errors (misses) of the ensemble of previous trees.

Gradient boosting combines these sequentially improved trees into an ensemble, where each tree corrects the errors of the ones before it. The "gradient" part comes from the use of gradient descent to minimize the loss function (the discrepancy between the predicted and actual values), guiding how the model adjusts for the errors with each new tree.

By the end of the process, the collective wisdom of all these incremental improvements (all the archers' adjusted shots) leads to a model that is much closer to making accurate predictions (hitting the target), even if each tree (archer) may not be perfect on its own. This iterative refinement makes gradient-boosted tree regression a powerful technique for reducing bias and variance, resulting in a highly accurate predictive model.

Applications

Let's review this type of regression's application:

- **Financial modeling**: Predicting stock prices, credit scoring, and risk management by analyzing numerous market factors
- **Real estate**: Estimating property values based on location, size, amenities, and market trends
- **Energy consumption forecasting**: Predicting future energy demand in households or industries to optimize supply and reduce costs

Benefits

The benefits are as follows:

- **High accuracy**: Often provides superior predictive accuracy compared to other algorithms
- **Flexibility**: Can handle various types of data, including numerical and categorical
- **Handling of complex non-linearities**: Effectively captures complex non-linear relationships between features and target

Limitations

Here are the limitations:

- **Overfitting risk**: Especially with noisy data and without proper regularization or if too many trees are used
- **Computationally intensive**: Training can be resource-heavy and time-consuming, particularly with large datasets
- **Model complexity**: Its sequential nature makes the model less interpretable than simpler models, complicating the process of understanding and explaining predictions.

Survival regression

Survival regression, particularly the Cox proportional hazards model, is a statistical approach that's used to investigate the effect of various factors on the time of a specific event, such as death, failure, or churn. Unlike standard regression, which deals with continuous or categorical outcomes, survival regression is designed to handle time-to-event data, which is often censored. Censoring occurs when the event of interest hasn't happened for some subjects during the study period, making their exact time-to-event unknown.

This model doesn't assume a particular statistical distribution for survival times. Instead, it models the hazard rate as a function of covariates. The hazard rate at any time is assumed to be a baseline hazard function multiplied by an exponential function of the covariates. This formulation allows for hazard ratios to be estimated, providing insights into the relative risk associated with covariates.

Imagine you're a gardener growing a variety of plants, each with different survival rates due to factors such as sunlight, water, soil type, and pest exposure. You're trying to predict how long each plant will thrive while considering these diverse conditions. In this scenario, each plant represents an individual in a dataset, and the factors affecting their growth are analogous to the covariates in survival analysis.

To tackle this, you decide to create a detailed chart (a "survival tree") for each type of plant, mapping out various conditions and their effects on plant longevity. For instance, you notice that plants with less than 6 hours of sunlight per day wilt earlier, so you create a branch in your chart for "sunlight < 6 hours" leading to a "shorter lifespan." If a plant gets more sunlight but is in clay-heavy soil, that might form another branch, showing these plants also have reduced survival times compared to those in loamy soil.

As you expand your chart, you create more branches based on your observations, effectively segmenting the plants into groups with similar survival expectations under specific conditions. This "tree" of decisions and outcomes helps you predict how long new plants might survive based on their conditions, guiding your gardening strategies to enhance their longevity.

This gardening strategy mirrors the concept of survival tree regression, where individuals (plants) are split into homogeneous groups (branches) based on their covariates (conditions such as sunlight and soil). This tree structure allows for an intuitive understanding of how different conditions interact to affect survival, much like how different gardening practices influence plant longevity. Just as you'd adjust your care strategies based on your chart to maximize plant survival, survival tree regression helps in understanding and strategizing interventions to improve outcomes based on the predictive model.

Applications

Survival regression has applications in the following fields:

- **Medical research**: Used extensively to compare the effectiveness of treatments, where the event might be death, relapse, or recovery

- **Customer analytics**: In analyzing customer churn, to understand how different factors contribute to the timing of customers leaving a service

- **Reliability engineering**: Applied to predict the lifespan of products or components, informing maintenance schedules and product design

Benefits

It brings the following benefits:

- **Censored data handling**: Effectively incorporates right-censored data, common in time-to-event studies, without discarding information

- **Risk assessment**: Provides a mechanism to assess the impact of various covariates on the risk (hazard) of an event occurring, offering valuable insights for risk management

- **Versatility**: Can be applied across various fields with time-to-event data, making it a versatile tool for statistical analysis

Limitations

However, it also has its pitfalls:

- **Proportional hazards assumption**: The assumption that the effects of covariates are multiplicative and constant over time may not hold in all situations, potentially biasing results

- **Interpretation challenges**: The complexity of the model and the concept of hazards can make interpreting and communicating results challenging, especially for non-specialist audiences

- **Data and event requirements**: Reliable analysis requires a sufficient number of events and a diverse range of covariate values, limiting its applicability in small or narrowly defined datasets

Factorization machine regressor

Factorization machine regressor is a versatile machine learning algorithm that captures interactions between variables in high-dimensional sparse datasets efficiently. It extends beyond the capabilities of matrix factorization by modeling all interactions in a dataset with a polynomial degree of 2, making it particularly powerful for tasks involving sparse data, such as recommendation systems, where traditional algorithms might struggle due to the vast number of possible feature interactions.

Imagine you're organizing a potluck dinner where each guest brings a dish, and you want to predict the overall success of the event based on the combination of dishes. Each guest has their preferences and cooking skills, and each dish has its unique flavor. The success of the potluck isn't just about the individual dishes but how well they complement each other to create a satisfying meal.

In this analogy, each guest and dish represent the features in your dataset, and the overall success of the potluck is the target variable you're trying to predict. Just as some dishes might go well together (such as wine and cheese) and others might not (such as ice cream and ketchup), the interactions between features in your data can significantly impact the outcome.

A factorization machine regressor is like a sophisticated potluck planner that not only considers the individual qualities of each guest's cooking skills and dish preferences but also how well each combination of dishes might contribute to the success of the meal. It does this by modeling not just the individual features but also the interactions between them, much like how you'd think about the balance of flavors, textures, and colors in the meal to ensure it's enjoyable.

By considering these interactions, the factorization machine regressor can make more accurate predictions about the potluck's success, just as careful planning and consideration of how dishes complement each other can lead to a more successful and enjoyable meal. This approach allows for a nuanced understanding of how different elements contribute to the outcome, capturing the complexity of real-world situations where the interplay between factors is often key to understanding the results.

Applications

Factorization machine regressor has the following applications:

- **Recommendation systems**: Factorization machines are particularly adept at handling the sparse data typical of user-item matrices in recommendation systems, providing personalized recommendations based on user behavior and item attributes
- **Click prediction**: In online advertising, factorization machines predict the likelihood of a user clicking on an advert by learning from user demographics, advert features, and historical click data, optimizing advert placements and targeting

- **Rating prediction**: They are used in rating systems, such as movie or product reviews, to predict user ratings based on past interactions and the characteristics of both users and items, enhancing content personalization.

Benefits

Here are the benefits:

- **Efficiency with sparse data**: Factorization machines excel in handling sparse datasets, capturing complex interactions with fewer parameters

- **Versatility**: They can be used for regression, classification, and ranking tasks, making them highly adaptable to various types of predictive modeling challenges

- **Scalability**: Despite modeling complex interactions, factorization machines are relatively scalable and can handle large datasets effectively

Limitations

Let's glance through the limitations as well:

- **Computational complexity**: Training factorization machines, especially on very large datasets, can be computationally intensive due to the need to model interactions between all features

- **Hyperparameter sensitivity**: The performance of factorization machines can be highly sensitive to the choice of hyperparameters, such as the dimensionality of the factorization, and thus require careful tuning

- **Data preprocessing**: Applying factorization machines effectively may require substantial feature engineering and preprocessing to structure the data appropriately so that it can capture meaningful interactions

Now that we understand several regression algorithms, let's learn how to evaluate the regression model's performance.

Evaluating the model's performance

In this section, we'll look at several evaluation metrics that can be used to evaluate the performance of the regression model.

Evaluating a regression model is a critical step in the modeling process, providing insights into how well the model performs in predicting continuous outcomes. This evaluation involves using various metrics to quantify the accuracy, reliability, and predictive power of the model. Evaluating regression models revolves around quantifying their accuracy and predictive performance, focusing on how well they estimate the continuous outcome variable based on the input features. Understanding these metrics and their implications can help in refining the model, selecting the most appropriate one for a given task, and ensuring that the predictions are useful for decision-making. Here's an in-depth look at the evaluation process and key metrics for regression models:

- **Mean squared error (MSE)**: MSE measures the average of the squares of the errors – that is, the average squared difference between the estimated values and the actual value. It gives a clear idea of how far the regression model's predictions are from the actual values. A lower MSE value indicates a better fit of the model to the data. MSE is sensitive to outliers because it squares the errors before averaging, which can significantly increase the error value when large errors are present.

- **Root mean squared error (RMSE)**: RMSE is the square root of the mean squared error. It is one of the most widely used metrics for regression analysis because taking the square root brings the scale of the errors back to the original units of the output variable, making interpretation easier. Like MSE, RMSE penalizes larger errors more heavily, but its units are easier to understand relative to the target variable.

- **Mean absolute error (MAE)**: MAE measures the average of the absolute errors between the predicted values and the actual observations. It provides a linear score that reflects the average magnitude of the errors in the predictions, without considering their direction. Unlike MSE and RMSE, MAE is not as sensitive to outliers, making it a useful metric when the dataset contains many anomalies or outliers.

- **R-squared (R^2)**: The coefficient of determination, or R^2, measures the variance proportion in the dependent variable that can be predicted by the independent variables. It explains how well the observed outcomes are replicated by the model, based on the proportion of total variation of outcomes explained by the model. R^2 values range from 0 to 1, where 1 indicates a perfect fit and 0 indicates that the model doesn't explain any variability in the target variable. While R^2 is widely used for its interpretability, it can sometimes be misleading, especially in models with many predictors or when used outside the context of the data from which the model was developed.

- **Adjusted R-squared**: Adjusted R-squared is a modified version of R^2 that has been adjusted for the number of predictors in the model. Unlike R^2, which can increase as more variables are added to the model, even if those variables are not significant, adjusted R-squared will only increase if a new variable improves the model more than would be expected by chance. This makes adjusted R-squared a more robust measure for comparing models with different numbers of independent variables.

Having looked at the different metrics we can use to evaluate a model, we also need to know how to select the right one for our model.

Selecting the evaluation metrics

Choosing the right evaluation metrics for a regression model depends on various factors related to the specific context of the problem, the nature of the data, and the objectives of the modeling task. Here are some considerations to help you select appropriate evaluation metrics for regression models:

- **Nature of the prediction task:**

  - **Accuracy importance:** If accurate predictions are crucial and errors have significant consequences, consider using metrics that heavily penalize larger errors, such as MSE or RMSE

  - **Relative error sensitivity:** If understanding errors in proportion to actual values is important (for example, in forecasting), metrics such as MAPE or **mean absolute scaled error (MASE)** might be more appropriate

- **Data characteristics:**

  - **Outlier sensitivity:** If your data is prone to outliers or has a skewed distribution, MAE can be a better choice because it's less sensitive to outliers compared to MSE or RMSE

  - **Scale dependency:** If the data spans several orders of magnitude, percentage-based errors such as MAPE can provide a scale-independent view of the model's performance

- **Model comparison and complexity:**

  - **Comparing models:** If you need to compare models with different numbers of predictors, adjusted R-squared can be useful as it adjusts for the number of predictors in the model. This is unlike R^2, which can increase with the number of predictors regardless of their contribution to the model.

  - **Model complexity:** For models that are at risk of overfitting, consider using metrics that can help in regularization, such as the **Akaike information criterion (AIC)** or **Bayesian information criterion (BIC)**, which penalize unnecessary complexity.

- **Interpretability and communication:**

  - **Stakeholder understanding:** Choose metrics that are easy for stakeholders to understand and interpret. MAE, for instance, is straightforward and can be easily communicated to non-technical audiences.

  - **Interpretation ease:** Metrics that are in the same units as the target variable, such as MAE and RMSE, can be easier to interpret in the context of the problem.

- **Specific objectives or constraints:**

  - **Business or domain-specific objectives:** The choice of metric can also depend on specific business objectives or constraints. For example, in finance, minimizing large errors might be more critical than in other domains, favoring the use of RMSE.

- **Cost-sensitive learning**: In some cases, the costs of different types of errors may not be the same. Custom loss functions or metrics that reflect the asymmetric cost of overpredictions versus underpredictions can be developed and used.

To conclude, there's no one-size-fits-all metric for regression models, and often, multiple metrics are used in conjunction to get a holistic view of the model's performance. The key is to match the metric to the specific needs of the task while considering the implications of errors, the nature of the data, and the requirements of the stakeholders involved. It's also essential to understand the limitations of each metric and avoid over-reliance on any single measure.

The next section will explain how to improve the model's performance.

Improving the model's performance

Improving the performance of a regression model involves a combination of data preprocessing, feature engineering, model selection, and tuning. Here are several strategies you can implement to enhance your model's predictive capabilities:

- **Data quality and preprocessing**:

  - **Handle missing data**: Use imputation techniques to fill in missing values or consider removing rows or columns with excessive missing data

  - **Outlier detection and treatment**: Identify and remove outliers that may skew the model or use robust methods to reduce their influence

  - **Feature scaling**: Normalize or standardize features, especially for algorithms sensitive to the scale of the data, such as gradient descent-based methods

- **Feature engineering**:

  - **Feature selection**: Use techniques such as backward elimination, forward selection, or regularization methods (lasso, ridge) to retain only the most relevant features.

  - **Feature transformation**: Apply transformations such as log, square root, or Box-Cox to make the data distribution more normal. This can be especially helpful for linear models.

  - **Feature construction**: Create new features from existing ones through domain knowledge or automated feature engineering tools. Polynomial features can model non-linear relationships.

- **Model complexity**:

  - **Simplification**: Simplify complex models to prevent overfitting. This might involve reducing the number of parameters, using simpler models, or applying regularization techniques.

  - **Ensembling**: Combine multiple models to improve predictions. Techniques such as bagging, boosting, and stacking can reduce variance and bias, leading to more robust models.

- **Algorithm selection**: Explore different regression algorithms and select the one that performs best for your specific dataset and problem. Each algorithm has its strengths and weaknesses depending on the nature of the data and the task.

- **Hyperparameter tuning**: Use grid search, random search, Bayesian optimization, or automated hyperparameter tuning tools to find the optimal settings for your model's hyperparameters.

- **Cross-validation**: Implement cross-validation techniques to assess how your model will generalize to an independent dataset. K-fold cross-validation is a standard approach that ensures every observation from the original dataset has the chance of appearing in the training and test set.

- **Incorporate domain knowledge**: Utilize domain knowledge to inform feature selection, model design, and interpretation. Domain-specific variables can significantly improve model performance.

- **Regularization techniques**: Apply regularization methods such as lasso (L1 regularization), ridge (L2 regularization), or elastic net to reduce overfitting by penalizing large coefficients.

Improving regression model performance is an iterative process that involves exploring various techniques and understanding the trade-offs between bias and variance. It's crucial to continuously monitor model performance and adjust as necessary based on new data, feedback, and changing conditions.

Next, let's review some model-specific strategies so that we can improve the model's performance:

- **Linear and generalized linear regression**:

  - Regularization (lasso, ridge, elastic net) can prevent overfitting by penalizing large coefficients

  - For generalized linear models, ensure the link function and the distribution family are appropriate for the data

- **Decision tree regression**:

  - Control the depth of the tree to prevent overfitting

  - Prune the tree to remove sections that provide little power to classify instances

- **Random forest and gradient-boosted tree regression**:

  - Tune the number of trees and their depth to balance model complexity and performance

  - Adjust the learning rate (for gradient-boosted tree regression) to control how quickly the model adapts to the "hard" observations

- **Survival regression**:

  - For Cox models, verify the proportional hazards assumption. If violated, consider stratified or time-varying coefficient models.

  - Use feature engineering to include time-dependent covariates if necessary.

- **Factorization machine regressor**:

    - Tune the dimensionality of the factorization space to capture sufficient interactions without overfitting

    - Regularize the model to manage the complexity brought by high-dimensional sparse data

Improving regression model performance is an iterative process that often requires balancing the trade-offs between model complexity, interpretability, and computational efficiency. Customizing these strategies to the specific characteristics of the dataset and the requirements of the task can lead to more accurate and reliable models.

Practical implementation

Now, let's delve into coding to understand the practical implementation of regression system.

About the dataset

The California Housing dataset is a popular dataset that's used in data science and machine learning to demonstrate algorithms and techniques for regression analysis. This dataset typically serves as a benchmark for predictive modeling tasks, where the goal is to forecast median house values in Californian districts based on various features. Originating from the 1990s, it was compiled for a study related to housing needs in California and has since become a staple example in machine learning communities, particularly for those starting with data science.

Context and features

The California Housing dataset consists of information from the 1990 US Census, aggregated to the block-group level, often referred to as "districts" in the context of this dataset. A block group is the smallest geographical unit for which the US Census Bureau publishes sample data, typically containing 600 to 3,000 people. The dataset encompasses several features that describe the characteristics of housing and the population within each district. These features include the following:

- **Median income**: The median income of households within a block, scaled and capped. This is a crucial predictor for housing prices as it reflects the purchasing power of the residents.

- **Housing median age**: The median age of houses in a block. This can indicate the level of newness or dilapidation of buildings.

- **Total rooms**: The total number of rooms (excluding bedrooms) in all houses within a block.

- **Total bedrooms**: The total number of bedrooms in all houses within a block.

- **Population**: The total number of people residing within a block.

- **Households**: The total number of households, defined as groups of people residing together, typically in a single dwelling unit.

- **Latitude and longitude**: Geographic coordinates of the block's centroid, useful for mapping and spatial analysis.

- **Target variable**: The primary target variable in the California Housing dataset is the median house value for each district. This variable represents the median value of owner-occupied homes, serving as the response variable that models the attempt to predict based on other features.

The California Housing dataset is widely used for teaching and testing regression models and is a type of predictive modeling technique that's used to understand relationships between independent variables (features) and a continuous dependent variable (target). In the context of this dataset, regression models are built to predict the median house value based on district-level characteristics.

Loading the California Housing data

The following lines of code mainly focus on loading the dataset, converting it into a pandas DataFrame, and then creating a Spark DataFrame from it:

```
from pyspark.sql import SparkSession
spark = SparkSession.builder.appName("Regression").getOrCreate()
from pyspark.ml.feature import VectorAssembler
from pyspark.ml.regression import (
    LinearRegression, GeneralizedLinearRegression,
    DecisionTreeRegressor, RandomForestRegressor, GBTRegressor)
from pyspark.ml import Pipeline
from pyspark.ml.evaluation import RegressionEvaluator
from sklearn.datasets import fetch_california_housing
housing = fetch_california_housing()
import pandas as pd
df = pd.DataFrame(
    data=housing.data,columns=housing.feature_names)
df['label'] = housing.target
df_data = spark.createDataFrame(df)
df_data.show(5)
```

The preceding code snippet will output the content in DataFrame format, as shown here:

```
+-------+--------+------------------+------------------+----------+------------------+--------+---------+-----+
|MedInc |HouseAge|          AveRooms|          AveBedrms|Population|          AveOccup|Latitude|Longitude|label|
+-------+--------+------------------+------------------+----------+------------------+--------+---------+-----+
|8.3252 |   41.0|6.984126984126984|1.0238095238095237|    322.0|2.5555555555555554|   37.88|  -122.23|4.526|
|8.3014 |   21.0|6.238137082601054|0.9718804920913884|   2401.0| 2.109841827768014|   37.86|  -122.22|3.585|
|7.2574 |   52.0|8.288135593220339| 1.073446327683616|    496.0|2.8022598870056497|   37.85|  -122.24|3.521|
|5.6431 |   52.0|5.8173515981735155|1.0730593607305936|    558.0| 2.547945205479452|   37.85|  -122.25|3.413|
|3.8462 |   52.0|6.281853281853282|1.0810810810810811|    565.0|2.1814671814671813|   37.85|  -122.25|3.422|
+-------+--------+------------------+------------------+----------+------------------+--------+---------+-----+
only showing top 5 rows
```

Figure 4.3 – The California Housing dataset – DataFrame output

Here's a breakdown of what each part of the code is doing:

- First, all the required libraries are imported.

- `housing = fetch_california_housing()`: This line fetches the California Housing dataset and stores it in the `housing` variable.

- `df = pd.DataFrame(data=housing.data, columns=housing.feature_names)`: This line creates a pandas DataFrame named `df` from `housing.data` that contains the features of the dataset. The column names for the DataFrame are set using `housing.feature_names`.

- `df['label'] = housing.target`: This line adds a new column named `label` to the DataFrame, `df`.

- `df_data = spark.createDataFrame(df)`: This line converts the pandas DataFrame, `df`, into a Spark DataFrame, `df_data`.

- `df_data.show(5)`: This line displays the first 5 rows of the Spark DataFrame, `df_data`.

Preparing the feature columns

The following code snippet prepares the dataset for machine learning by transforming the feature columns into a format compatible with Spark MLlib's algorithms, allowing the models to be trained on large-scale data:

```
feature_cols = df_data.columns[:-1]
assembler = VectorAssembler(
    inputCols=feature_cols, outputCol="features")
df_data = assembler.transform(df_data)
df_data.show(5,100)
```

This time, we get the following output:

MedInc	HouseAge	AveRooms	AveBedrms	Population	AveOccup	Latitude	Longitude	label	features
8.3252	41.0	6.984126984126984	1.0238095238095237	322.0	2.5555555555555554	37.88	-122.23	4.526	[8.3252,41.0,6.98...]
8.3014	21.0	6.238137082601054	0.9718804920913884	2401.0	2.109841827768014	37.86	-122.22	3.585	[8.3014,21.0,6.23...]
7.2574	52.0	8.288135593220339	1.073446327683616	496.0	2.8022598870056497	37.85	-122.24	3.521	[7.2574,52.0,8.28...]
5.6431	52.0	5.8173515981735155	1.0730593607305936	558.0	2.547945205479452	37.85	-122.25	3.413	[5.6431,52.0,5.81...]
3.8462	52.0	6.281853281853282	1.0810810810810811	565.0	2.1814671814671813	37.85	-122.25	3.422	[3.8462,52.0,6.28...]

only showing top 5 rows

Figure 4.4 – DataFrame output showing features

Here's a breakdown of the code:

- `feature_cols = df_data.columns[:-1]`: This line selects all columns of the `df_data` DataFrame except the last one and stores them in the `feature_cols` list.

- `assembler = VectorAssembler(inputCols=feature_cols, outputCol="features")`: Here, an instance of the `VectorAssembler` transformer is created. Note that `VectorAssembler` is a feature transformer from PySpark MLlib that merges multiple columns of a DataFrame into a single vector column. This is often a necessary step because many machine learning algorithms in Spark's MLlib expect input data to be in this format:

  - The `inputCols` parameter is set to `feature_cols`, which means that the columns listed in `feature_cols` will be combined into a single vector

  - The `outputCol` parameter specifies the name of the new column (`features`) that will contain the merged vector

- `df_data = assembler.transform(df_data)`: This line applies the `VectorAssembler` transformer to `df_data`. The transform method of `VectorAssembler` takes the original `df_data` DataFrame, combines the specified `inputCols` into a single vector, and appends this vector as a new column named `features` to the DataFrame.

Feature scaling and normalization

Feature scaling and normalization are important preprocessing steps in the data preparation phase of machine learning and data modeling. These techniques adjust the scale or distribution of features (variables) in your data, improving model performance, ensuring faster convergence, and providing balanced regularization.

The following code snippet uses `StandardScaler`, a feature transformation tool used for scaling and normalizing features in data preprocessing for machine learning models:

```
from pyspark.ml.feature import StandardScaler
scaler = StandardScaler(inputCol="features",
    outputCol="scaledFeatures", withStd=True, withMean=True)
scalerModel = scaler.fit(df_data)
scaledData = scalerModel.transform(df_data)
```

Here's a breakdown of the code:

- `scaler = StandardScaler(inputCol="features", outputCol="scaledFeatures", withStd=True, withMean=True)`: This line initializes a `StandardScaler` object.

  - First, `inputCol="features"` specifies the name of the input column that contains the features vector you want to scale.

  - Then, `outputCol="scaledFeatures"` specifies the name of the new column where the scaled features will be stored.

- Next, `withStd=True` indicates that the scaler will scale the data to unit standard deviation.

- Finally, `withMean=True` indicates that the scaler will center the data with the mean before scaling. This means it will subtract the mean of each feature so that it's centered around zero.

- `scalerModel = scaler.fit(df_data)`: This line fits `StandardScaler` to the data (`df_data`), which computes the mean and standard deviation for each feature in the dataset. The result is a model (`scalerModel`) that contains these statistics.

- `scaledData = scalerModel.transform(df_data)`: This line applies the scaling transformation to the original data (`df_data`) using `scalerModel`. It scales the features in `inputCol` according to the means and standard deviations that were computed during the fitting stage.

The scaled features are stored in a new column specified by `outputCol` – in this case, `scaledFeatures`.

Splitting the dataset into training and testing sets

Splitting data is a fundamental practice in machine learning for validating models. It helps prevent overfitting, where a model might perform well on the training data but poorly on new, unseen data:

```
(trainingData, testData) = scaledData.randomSplit([0.8, 0.2])
```

The previous code snippet splits the `df_data` DataFrame into two separate DataFrames: `trainingData` and `testData`. This is achieved using the `randomSplit` method provided by Apache Spark, which partitions the DataFrame into subsets randomly.

Initializing regression models

Initializing machine learning models, such as the ones in our code snippet, is a crucial step in the machine learning workflow.

The following code snippet initializes five different regression models from Apache Spark's MLlib, each designed for different types of regression tasks. All these models are configured with a specified features column named `features` and a label column named `label`. In Spark MLlib, the `featuresCol` parameter is used to specify the input column that contains feature vectors, and the `labelCol` parameter specifies the column containing the target variable:

```
lr = LinearRegression(featuresCol="features", labelCol="label")
glr = GeneralizedLinearRegression(featuresCol="features",
    labelCol="label")
dt = DecisionTreeRegressor(featuresCol="features", labelCol="label")
rf = RandomForestRegressor(featuresCol="features", labelCol="label")
gbt = GBTRegressor(featuresCol="features", labelCol="label")
```

Each of these models is suitable for different regression tasks, depending on the nature of the data and the specific requirements of the problem.

Defining a pipeline for each regression algorithm

Pipelines in Spark MLlib are powerful tools for building machine learning workflows. They allow you to chain multiple transformation and model training stages, ensuring that all the steps are executed in the correct order. This is particularly useful for data preprocessing, feature transformation, and model training and evaluation to be bundled together into a single, reusable workflow. Pipelines also help in maintaining clean code and facilitate the deployment and reuse of machine learning models.

The following code snippet creates five separate pipelines using Apache Spark's Pipeline API, with each pipeline containing one of the machine learning regression models shown previously:

```
pipeline_lr = Pipeline(stages=[lr])
pipeline_glr = Pipeline(stages=[glr])
pipeline_dt = Pipeline(stages=[dt])
pipeline_rf = Pipeline(stages=[rf])
pipeline_gbt = Pipeline(stages=[gbt])
```

In this case, the pipeline contains only one stage: the regression models themselves.

Fitting the pipelines

The following code snippet trains the machine learning models on a dataset. Each line of code fits a different regression model to the training data, creating trained models that can be used for predictions:

```
model_lr = pipeline_lr.fit(trainingData)
model_glr = pipeline_glr.fit(trainingData)
model_dt = pipeline_dt.fit(trainingData)
model_rf = pipeline_rf.fit(trainingData)
model_gbt = pipeline_gbt.fit(trainingData)
```

This approach allows different regression models' performance to be evaluated and compared on the same training dataset.

Making predictions

The following code snippet takes the model that's been trained on the training data and applies it to unseen test data to evaluate the model's performance. The transform method in Spark MLlib is used for this purpose, and it adds a column (typically named prediction) to the input DataFrame that contains the predicted values for each row:

```
predictions_lr = model_lr.transform(testData)
predictions_glr = model_glr.transform(testData)
predictions_dt = model_dt.transform(testData)
predictions_rf = model_rf.transform(testData)
predictions_gbt = model_gbt.transform(testData)
```

We'll get the following output:

```
+------+--------+------------------+------------------+----------+------------------+--------+--------+------+--------------------+------------------+
|MedInc|HouseAge|          AveRooms|          AveBedrms|Population|          AveOccup|Latitude|Longitude|label|            features|        prediction|
+------+--------+------------------+------------------+----------+------------------+--------+--------+------+--------------------+------------------+
| 0.536|    46.0|3.142857142857143|1.0476190476190477|      37.0|1.7619047619047619|   38.02| -121.84|0.875|[0.536,46.0,3.142...|0.9702947933237596|
|0.6991|    26.0|2.669021190716448| 1.014127144298688|    1660.0|1.6750756811301715|   36.72| -119.73|0.895|[0.6991,26.0,2.66...| 0.511258901436527|
| 0.716|    39.0|4.730769230769231|1.0961538461538463|     316.0| 6.076923076923077|   37.96| -122.36|1.042|[0.716,39.0,4.730...|1.0889862873494138|
|0.8026|    23.0|5.369230769230769|1.1507692307692308|    1054.0| 3.243076923076923|   37.81| -122.29|1.125|[0.8026,23.0,5.36...|0.9874381691604412|
|0.8639|    28.0|4.289377289377289|1.0952380952380953|    1193.0|4.3699633699633695|   35.38| -118.98|0.494|[0.8639,28.0,4.28...|0.7235862206275385|
+------+--------+------------------+------------------+----------+------------------+--------+--------+------+--------------------+------------------+
only showing top 5 rows
```

Figure 4.5 – DataFrame output with prediction

Evaluating the models

The following code snippet evaluates the performance of different regression models on the test dataset using the RMSE metric:

```
evaluator = RegressionEvaluator(labelCol="label",
    predictionCol="prediction", metricName="rmse")
rmse_lr = evaluator.evaluate(predictions_lr)
rmse_glr = evaluator.evaluate(predictions_glr)
rmse_dt = evaluator.evaluate(predictions_dt)
rmse_rf = evaluator.evaluate(predictions_rf)
rmse_gbt = evaluator.evaluate(predictions_gbt)

print("Linear Regression RMSE:", rmse_lr)
print("General Linear Regression RMSE:", rmse_glr)
print("Decision Tree Regression RMSE:", rmse_dt)
print("Random Forest Regression RMSE:", rmse_rf)
print("Gradient Boosted Tree Regression RMSE:", rmse_gbt)
```

The output is as follows:

```
Linear Regression RMSE: 0.7187607164753942
General Linear Regression RMSE: 0.7187607164753942
Decision Tree Regression RMSE: 0.7313670844226426
Random Forest Regression RMSE: 0.6845742384733482
Gradient Boosted Tree Regression RMSE: 0.5693840814678601
```

Cross-validation and hyperparameter fine-tuning

Each `paramGrid` object is built using `ParamGridBuilder` and specifies a set of parameters and their values to be considered during the tuning process of the respective machine learning models. This process is critical for finding the optimal model parameters that will yield the best performance:

```
from pyspark.ml.tuning import ParamGridBuilder, CrossValidator
paramGrid_lr = ParamGridBuilder() \
```

```
        .addGrid(lr.regParam, [0.1, 0.01]) \
        .addGrid(lr.elasticNetParam, [0.0, 0.5, 1.0]) \
        .build()

paramGrid_glr = ParamGridBuilder() \
        .addGrid(glr.regParam, [0.1, 0.01]) \
        .addGrid(glr.maxIter, [10, 20]) \
        .build()

paramGrid_dt = ParamGridBuilder() \
        .addGrid(dt.maxDepth, [5, 10, 20]) \
        .addGrid(dt.maxBins, [16, 32, 64]) \
        .build()

paramGrid_rf = ParamGridBuilder() \
        .addGrid(rf.numTrees, [5, 10, 20]) \
        .addGrid(rf.maxDepth, [4, 6, 8]) \
        .build()

paramGrid_gbt = ParamGridBuilder() \
        .addGrid(gbt.maxDepth, [4, 6, 8]) \
        .addGrid(gbt.maxIter, [2, 4, 6]) \
        .build()
```

Here's a breakdown of the code for linear regression:

- Import statement: Here, `ParamGridBuilder` is used to build a grid of parameters to search over during model tuning, and `CrossValidator` (though not used in this snippet) is typically used to perform cross-validation over a set of parameters.

- Parameter grid for linear regression (`paramGrid_lr`):

  - `.addGrid(lr.regParam, [0.1, 0.01])`: This adds a grid of regularization parameters (`regParam`) to the parameter grid for a linear regression model. The regularization parameters that are being considered are 0.1 and 0.01.

  - `.addGrid(lr.elasticNetParam, [0.0, 0.5, 1.0])`: This adds a grid of elastic net mixing parameters (`elasticNetParam`) to the parameter grid. The parameters being considered are 0.0 (L2 regularization only), 0.5 (equal mix of L1 and L2), and 1.0 (L1 regularization only).

Defining evaluator and CrossValidator

The following code snippet sets up cross-validation for different machine learning algorithms. Cross-validation is a technique that's used to evaluate a machine learning model's predictive performance and ensure that it does not overfit the training data:

```
# Define evaluator
evaluator = RegressionEvaluator(labelCol="label"
    predictionCol="prediction", metricName="rmse")
```

The following code snippet is the cross-validator for linear regression:

```
# Define CrossValidator for each algorithm
cv_lr = CrossValidator(estimator=pipeline_lr,
                       estimatorParamMaps=paramGrid_lr,
                       evaluator=evaluator,
                       numFolds=3)
```

The following code snippet is the cross-validator for generalized linear regression:

```
cv_glr = CrossValidator(estimator=pipeline_glr,
                        estimatorParamMaps=paramGrid_glr,
                        evaluator=evaluator,
                        numFolds=3)
```

The cross-validator code snippet for the decision tree is as follows:

```
cv_dt = CrossValidator(estimator=pipeline_dt,
                       estimatorParamMaps=paramGrid_dt,
                       evaluator=evaluator,
                       numFolds=3)
```

The following code snippet is the cross-validator for random forest:

```
cv_rf = CrossValidator(estimator=pipeline_rf,
                       estimatorParamMaps=paramGrid_rf,
                       evaluator=evaluator,
                       numFolds=3)
```

The following code snippet is the cross-validator for gradient-boosted trees:

```
cv_gbt = CrossValidator(estimator=pipeline_gbt,
                        estimatorParamMaps=paramGrid_gbt,
                        evaluator=evaluator,
                        numFolds=3)
```

Let's break down the linear regression code block. The CrossValidator (estimator=pipeline_lr, estimatorParamMaps=paramGrid_lr, evaluator=evaluator, numFolds=3) line creates a cross-validator for a linear regression model. It specifies that the logistic regression model is part of pipeline_lr, uses the parameter grid defined as paramGrid_lr for hyperparameter tuning, evaluates models using the previously defined evaluator, and performs three-fold cross-validation (the dataset is divided into three parts, with each part being used as a test set once).

Fitting the CrossValidator instances

The following code snippet executes the fitting process for each defined CrossValidator instance on a training data dataset. This process involves training and evaluating the machine learning models using cross-validation with the specified parameter grids and pipelines for different algorithms:

```
cvModel_lr = cv_lr.fit(trainingData)
cvModel_glr = cv_glr.fit(trainingData)
cvModel_dt = cv_dt.fit(trainingData)
cvModel_rf = cv_rf.fit(trainingData)
cvModel_gbt = cv_gbt.fit(trainingData)
```

Here, cvModel_lr = cv_lr.fit(trainingData) runs the cross-validation process for the linear regression model pipeline (cv_lr), which includes training and validating the model across different parameter combinations specified in paramGrid_lr (from the previous code snippet provided). The fitting process uses the trainingData dataset. The best model that's found through this cross-validation process is stored in cvModel_lr.

Making the predictions

The following code snippet applies trained models to a test dataset to generate predictions:

```
predictions_lr = cvModel_lr.transform(testData)
predictions_glr = cvModel_glr.transform(testData)
predictions_dt = cvModel_dt.transform(testData)
predictions_rf = cvModel_rf.transform(testData)
predictions_gbt = cvModel_gbt.transform(testData)
```

Here, predictions_lr = cvModel_lr.transform(testData) uses the linear regression model (cvModel_lr), which was trained using cross-validation on the training dataset, to predict the outcomes for testData. The transform method applies the trained model to the test dataset, generating predictions stored in predictions_lr.

Evaluating the models

The following code evaluates the performance of various machine learning models on a test dataset using the RMSE:

```
rmse_lr = evaluator.evaluate(predictions_lr)
rmse_glr = evaluator.evaluate(predictions_glr)
rmse_dt = evaluator.evaluate(predictions_dt)
rmse_rf = evaluator.evaluate(predictions_rf)
rmse_gbt = evaluator.evaluate(predictions_gbt)

# Print the RMSE of each model
print("Linear Regression RMSE:", rmse_lr)
print("Linear Regression RMSE:", rmse_glr)
print("Decision Tree Regression RMSE:", rmse_dt)
print("Random Forest Regression RMSE:", rmse_rf)
print("Gradient Boosted Tree Regression RMSE:", rmse_gbt)
```

This will generate the output shown here:

```
Linear Regression RMSE: 0.7154279207799202
General Linear Regression RMSE: 0.7154279207799202
Decision Tree Regression RMSE: 0.7062000747791418
Random Forest Regression RMSE: 0.6589518171055013
Gradient Boosted Tree Regression RMSE: 0.5691318056290677
```

Figure 4.6 – RMSE output

With that, we've come to the end of this chapter.

Summary

Regression analysis plays a crucial role in supervised learning, enabling us to understand relationships between variables and make informed predictions. This chapter delved into various regression algorithms, including linear regression, generalized linear regression, decision tree regression, random forest regression, gradient-boosted tree regression, survival regression, and factorization machine regressor. Each algorithm offers unique benefits and applications across different fields, such as business, medicine, agriculture, engineering, education, and science.

Understanding the pitfalls and assumptions of regression techniques is vital to ensure accurate and reliable results. Evaluating model performance using metrics such as MSE, RMSE, MAE, and R^2 helps in selecting the best model for a given task. By following best practices in data preprocessing, feature engineering, and model tuning, you can significantly improve regression model performance, leading to more accurate predictions and better decision-making.

In the next chapter, we'll look into supervised classification algorithms.

5

Building a
Classification System

In this chapter, we'll look into classification, a type of supervised learning and a cornerstone of machine learning that has revolutionized the way we interpret and categorize data. As discussed in the previous chapter, supervised learning, at its core, involves training a model on a labeled dataset where each example is paired with the correct output, enabling the model to learn the mapping from inputs to outputs. Within this vast realm, classification stands out as a critical task that assigns inputs to two or more classes, making it indispensable for a myriad of applications, from email filtering to medical diagnosis.

We'll begin by laying out the foundational concepts of supervised learning. With a strong theoretical grounding, we'll then delve into the mechanics of various classification algorithms, such as decision trees, **support vector machine (SVM)**, K-nearest neighbors, and so on, each with its unique strengths and applications.

To understand these concepts, we'll explore practical case studies and real-world examples that showcase the transformative power of classification in sectors such as finance, healthcare, and cybersecurity. Through hands-on exercises, you will gain the skills to preprocess data, select appropriate models, and fine-tune parameters to optimize performance.

Moreover, we'll also look into the critical aspects of evaluating classifier performance, introducing metrics such as accuracy, precision, recall, and the confusion matrix, ensuring you have the tools to assess and improve your models rigorously. Finally, we'll look into improving the model's performance. We'll learn about data quality, handling data imbalance, hyperparameters tuning, regularization, and so on.

In this chapter, we'll cover the following topics:

- Learning about classification
- Learning about classification algorithms
- Evaluating the model's performance
- Improving the model's performance

By the end of this chapter, you will not only have a comprehensive understanding of classification but also the practical experience to apply these techniques to solve complex problems, marking a significant milestone in your journey through the fascinating world of machine learning.

Technical requirements

You can find the code files for this chapter on GitHub at `https://github.com/PacktPublishing/Apache-Spark-for-Machine-Learning/tree/main/Chapter05`.

Learning about classification

This section will provide a general overview of classification, including its benefits, drawbacks, and practical use cases.

Classification overview

Classification, a fundamental task within supervised learning in machine learning, involves categorizing data points into predefined classes or groups. This process is important for making sense of diverse datasets and has widespread applications across various domains, including spam detection in emails, sentiment analysis in social media, and disease diagnosis in healthcare.

At its core, classification relies on a labeled dataset, where each instance is associated with a known category. The goal of a classification algorithm is to analyze this dataset and develop a model capable of predicting the class for new, unseen instances based on their attributes. These models map input features to output categories and are trained through algorithms that adjust their parameters to minimize the difference between predicted and actual class labels on the training data.

The following figure provides an example of email spam detection using classification:

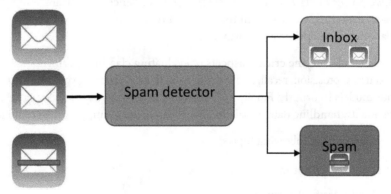

Figure 5.1 – Classification example

Let's explore classification further.

When to use the classification technique

Classification should be used when you need to predict which category or class a particular data point belongs to. It's most appropriate in scenarios where the output variable is categorical, meaning it represents discrete labels or categories. Here are some common situations where classification is the ideal choice:

- **Binary decisions**: Whenever there are two possible categories, such as yes/no, true/false, spam/not spam, or pass/fail scenarios. An example of this would be determining whether a loan application should be approved or denied.

- **Multi-class categorization**: When there are more than two categories and each instance needs to be classified into one of these. An example would be categorizing news articles into topics such as sports, politics, or technology.

Some use cases of classification in machine learning

Classification in machine learning offers a wide array of practical applications across various industries and domains. Here are some notable use cases:

- **Email spam detection**: Machine learning models are trained on features such as email content, frequency of certain words, and sender information to classify incoming emails accurately.

- **Customer segmentation**: Businesses utilize classification to group customers into distinct segments based on purchasing behavior, demographics, and preferences.

- **Disease diagnosis**: In healthcare, classification models are used to diagnose diseases by analyzing patient data, including symptoms, test results, and medical imaging.

- **Fraud detection**: Financial institutions apply classification to identify fraudulent transactions. By learning from historical transaction data, models can flag transactions that deviate from typical patterns, helping to prevent fraud in credit card transactions, insurance claims, and more.

- **Sentiment analysis**: In natural language processing, classification is used to determine the sentiment of text data, such as reviews, social media posts, and customer feedback.

- **Credit scoring**: By analyzing applicants' financial histories, demographics, and other relevant features, models can classify individuals into different risk categories, aiding in the decision-making process for loan approvals.

Drawbacks of classification techniques

While classification techniques are powerful tools in machine learning, they come with their own set of drawbacks and challenges that can affect performance, applicability, and ethical considerations. Understanding these drawbacks is crucial for effective model development and application:

- **Overfitting**: One of the most common issues, overfitting occurs when the model learns the training data too well, including its noise and outliers, making it perform poorly on unseen data.

- **Imbalanced data**: When the classes in the training data are not represented equally, models tend to be biased toward the majority class, leading to poor performance in the minority class.

- **Bias and fairness**: Classification models can inadvertently learn and perpetuate biases present in the training data, leading to unfair or discriminatory outcomes.

- **Simplicity versus complexity**: Simple models may fail to capture complex patterns in data (underfitting), while overly complex models may become difficult to interpret and computationally expensive. Striking the right balance is key.

- **Feature selection and engineering**: The choice and quality of input features significantly affect model performance. Irrelevant or redundant features can confuse the model, whereas important features may be overlooked or encoded improperly.

- **Data quality and availability**: Poor data quality, including missing values, outliers, and errors, can lead to inaccurate models.

Let's learn more about classification algorithms.

Learning about classification algorithms

Now, let's understand various classification algorithms. Classification algorithms are fundamental tools in machine learning. They enable us to categorize data points into predefined classes or labels. Whether you're solving a spam detection problem, predicting customer churn, or diagnosing diseases, understanding classification algorithms is essential. In this section, we'll go through the classification algorithms available in Apache Spark.

Logistic regression classification

Logistic regression is a predictive analysis algorithm that's used in the field of statistics and machine learning for binary classification problems. It models the probability that a given input belongs to a particular category, typically returning an output between 0 and 1 through the logistic function. This method is widely favored for cases where the outcome needs to be divided into two distinct classes, such as "yes" versus "no" or "success" versus "failure."

Unlike linear regression, which predicts continuous outcomes, logistic regression is designed for categorical outcomes. It estimates the odds of an event occurring by applying the logistic function to a linear combination of the input features, ensuring that the output probabilities are confined within the 0 to 1 range. This is particularly useful for problems where you're more interested in the likelihood of outcomes rather than their numerical values.

Imagine you're the coach of a soccer team and you're trying to predict whether your team will win the next match. You consider various factors: the team's overall health, the number of practice hours, the morale of the team, and whether the game is at home or away. In this scenario, each factor can be thought of as a feature in a logistic regression model, and the outcome (win or not win) is binary.

Logistic regression works like you weighing the importance of each factor based on past games. You might notice that games played at home with high team morale significantly increase the chances of winning. So, in your mental model, you give more "weight" to these factors. However, not every factor is straightforward; the relationship isn't purely additive. Increasing practice hours from 1 hour to 2 might significantly boost the likelihood of winning but increasing from 10 hours to 11 might not make much difference. Logistic regression captures this through the logistic function, which scales the weighted sum of the features to a probability (between 0 and 1).

When you predict the outcome of the next game, you're essentially calculating the odds of winning based on your weighted factors, much like logistic regression computes the probability of the target class (winning) based on the input features. If the computed probability is above a certain threshold (say, 0.5), you'd predict a win; otherwise, you'd brace for a loss. This process mirrors how logistic regression classifies instances based on the predicted probabilities derived from the model.

The following figure explains the working of the logistics regression classifier by showing several components that can yield weight, such as the input data and activation function:

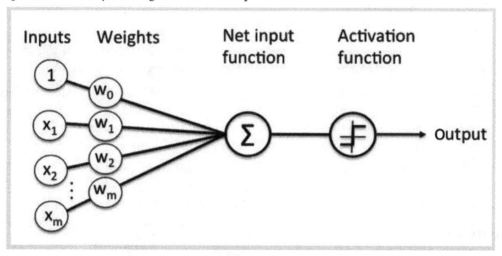

Figure 5.2 – Working mechanism of the logistics regression classifier

The input data represents the characteristics of the data points. The weights determine the importance of each feature in predicting the output. Finally, the activation function maps the linear combination of features and weights to a probability value between 0 and 1.

The following figure shows a sigmoid curve, which can be used to map all the data point values between 0 and 1. The labels are assigned based on the threshold value.

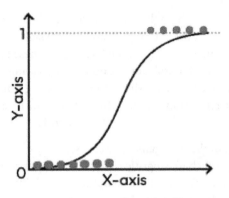

Figure 5.3 – Sigmoid curve

Applications

Here are three notable applications:

- **Medical diagnosis**: In healthcare, logistic regression is extensively used to predict the presence or absence of diseases, such as diabetes or cancer, based on patient data such as age, weight, genetic markers, and blood parameters. It helps in early detection and preventive healthcare planning.

- **Credit scoring**: Financial institutions leverage logistic regression to evaluate the creditworthiness of individuals. By analyzing various factors, such as income level, credit history, and employment status, the model predicts the likelihood of a borrower defaulting on a loan, aiding in risk assessment and decision-making.

- **Customer churn prediction**: Businesses use logistic regression to forecast customer retention, identifying which customers are likely to discontinue a service or product. This insight allows companies to implement targeted retention strategies and improve customer satisfaction.

Benefits

The top three benefits of logistics regression are as follows:

- **Interpretability**: One of the key strengths of logistic regression is its interpretability; the model's coefficients can be directly associated with the odds of the dependent variable, providing clear insights into how each feature influences the outcome.

- **Simple and efficient**: Logistic regression is computationally straightforward and efficient, making it a quick solution for binary classification problems, especially with smaller datasets or when a baseline model is needed.

- **Probabilistic predictions**: The algorithm provides not just classifications but also the probabilities of those classifications, offering a quantified confidence level in its predictions. This can be crucial for decision-making processes.

Limitations

Among the limitations of logistics regression, the following are the top three that often pose significant challenges:

- **Linearity assumption**: Logistic regression expects a linear relationship between the independent variables and the dependent variable, which might not always accurately represent real-world complexities

- **Limited to binary outcomes**: While it excels in binary classification tasks, logistic regression is not suitable for predicting continuous outcomes or multi-class classification problems without modifications

- **Overfitting risk**: In scenarios with a large number of features or insufficient data, logistic regression is prone to overfitting, leading to models that perform well on training data but generalize poorly to new, unseen data

Despite its limitations, logistic regression remains a popular algorithm for statisticians and data scientists for its simplicity, interpretability, and efficacy in binary classification tasks across diverse applications.

Decision tree classifier

A decision tree classifier is a non-parametric supervised learning method that's used for classification tasks. It resembles a tree structure, where each internal node represents a "test" on an attribute, each branch represents the outcome of the test, and each leaf node represents a class label. The paths from root to leaf represent classification rules.

In essence, a decision tree splits the dataset into subsets based on the most significant attributes, making the decision-making process transparent. This process is recursive and is known as **recursive partitioning**. The tree is built by selecting the attribute that best separates the data into classes, using measures such as Gini impurity or information gain in a process called **splitting criteria**. The process continues until it meets a stopping criterion, such as the maximum depth of the tree or no further improvement in the splitting criteria.

Imagine you're planning an outdoor picnic and need to decide whether to go based on the day's weather conditions. You create a simple decision-making process, much like a flowchart, to help you decide.

First, you consider the weather: if it's raining, you decide against the picnic. If it's not raining, you move to the next consideration: the temperature. If it's too cold, you decide against it; if it's warm, you then consider the wind speed. If it's too windy, you might decide it's not a good day for a picnic; if there's little to no wind, you decide to go ahead with the picnic.

This process, where you make a series of binary decisions (yes/no) based on certain conditions, closely resembles how a decision tree classifier works. Each "node" in your decision process (rain, temperature, and wind speed) is analogous to the nodes in a decision tree where a feature is evaluated. The final "decisions" (go on a picnic or not) are akin to the leaves of the decision tree, representing the outcome or classification based on the input features.

Just like your picnic decision process becomes more refined with each condition considered, a decision tree classifier iteratively splits the data into increasingly homogenous groups (or nodes) based on the features that most effectively classify the data until it reaches a decision (or leaf node).

The following figure shows an example of the decision tree classifier:

Is the person fit or unfit?

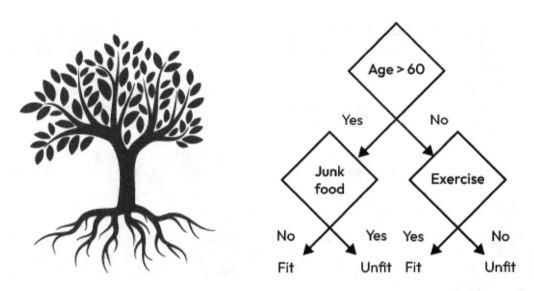

Figure 5.4 – An example of the decision tree classifier

Applications

Here are three notable applications of the decision tree classifier:

- **Customer segmentation**: Decision trees are widely used in marketing and business strategies for segmenting customers into distinct groups based on specific criteria, such as demographics, purchasing behavior, and product preferences. This helps businesses tailor their marketing efforts and product offerings to different customer segments.

- **Fraud detection**: In the banking and finance sector, decision trees are employed to identify suspicious activities by analyzing patterns in transaction data. The tree's branches might represent different aspects of a transaction, such as amount, location, and frequency, aiding in the efficient detection of fraudulent transactions.

- **Medical diagnosis**: Decision trees can assist healthcare professionals in diagnosing diseases by analyzing patients' symptoms and test results. Each node in the tree might represent a symptom or test result, and the leaves might represent potential diagnoses, guiding the diagnostic process.

Benefits

The top three benefits of decision tree classifier are as follows:

- **Interpretability**: One of the primary advantages of decision trees is their simplicity and ease of interpretation. The tree structure is intuitive and can be easily visualized, making the decision-making process transparent and understandable even to non-experts.

- **Handling of non-linear relationships**: Decision trees can capture non-linear relationships between features without the need for transformation, making them suitable for complex datasets where linear discriminants do not perform well.

- **No need for data pre-processing**: Unlike many other algorithms, decision trees do not require the data to be normalized or scaled, and they can handle both numerical and categorical data effectively.

Limitations

Among the limitations of the decision tree classifier, the top three that often pose significant challenges are as follows:

- **Overfitting**: Without proper tuning, decision trees can create overly complex trees that do not generalize well to unseen data. Overfitting can be mitigated by pruning the tree, setting a maximum depth, or requiring a minimum number of samples per leaf.

- **Instability**: Decision trees can be sensitive to small variations in the data, leading to different tree structures. This instability can be reduced by using ensemble methods such as random forest, which builds multiple trees and aggregates their predictions.

- **Bias toward dominant classes**: Decision trees can be biased toward classes with a larger number of instances, leading to biased predictions. Balancing the dataset or adjusting class weights can help address this issue.

Despite these limitations, decision trees remain popular due to their simplicity, interpretability, and flexibility, making them a valuable tool for a wide range of classification tasks.

Random forest classifier

A random forest classifier is an ensemble learning method that's widely regarded for its accuracy and robustness in handling various classification tasks. It operates by constructing multiple decision trees during the training phase and outputting the class that is the majority class of the individual trees. Random forest helps reduce overfitting to their training set, making it more generalizable.

The algorithm introduces randomness in two ways. First, each tree in the forest is built from a random sample of the data and drawn with replacement, known as **bootstrapping**. Second, at each node, a random subset of features is considered for splitting. These elements of randomness ensure that the trees are de-correlated, reducing the variance without increasing the bias significantly, leading to a model that performs well on unseen data.

Imagine you're trying to decide on the perfect movie to watch with a group of friends on movie night. Everyone has different tastes and preferences, making it a complex decision. To solve this, you decide to consult several of your cinephile friends, each with their unique taste and expertise in different movie genres.

Each friend acts as an "expert" and makes a movie recommendation based on certain criteria: one might suggest a movie based on its director, another might consider the lead actors, while another might focus on the genre or the movie's awards. This is akin to individual decision trees in a random forest, where each tree makes a prediction based on a subset of features (criteria).

Instead of relying on a single friend's recommendation (which could be biased or limited), you decide to take a vote among all your cinephile friends, with the most recommended movie being chosen for movie night. This collective decision-making process mirrors how a random forest classifier aggregates predictions from multiple decision trees through a majority vote or averaging process, leading to a more robust and accurate prediction than any single decision tree could provide.

Just as your group of friends brings diverse perspectives and expertise to make a well-rounded movie choice, a random forest as shown combines the predictions of many decision trees, each trained on different subsets of the data and features, reducing the risk of bias or overfitting, and enhancing the overall decision-making process:

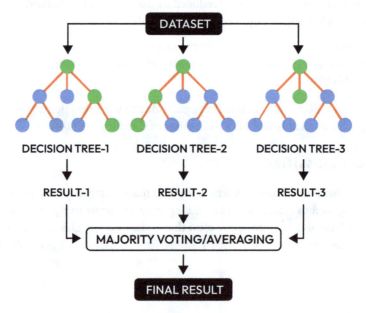

Figure 5.5 – Working mechanism of the random forest classifier

Applications

Here are three notable applications of the random forest classifier:

- **Biomedical engineering**: Random forests are used in gene expression data analysis for classifying diseases and identifying important biomarkers, aiding in diagnostics and personalized medicine

- **Banking sector**: In finance, random forest classifiers are instrumental in detecting fraudulent activities as they can analyze transaction patterns and identify anomalies that deviate from the norm

- **E-commerce**: They help in recommendation systems by classifying products based on user preferences and browsing history, enhancing the shopping experience through personalized recommendations

Benefits

The top three benefits of random forest classifiers are as follows:

- **Accuracy**: One of the most significant advantages of random forest classifiers is their high accuracy. The ensemble approach, through the process of aggregating predictions from multiple trees, tends to yield more accurate results than individual decision trees.

- **Robustness to overfitting**: Due to the averaging of multiple trees, random forests are much less prone to overfitting compared to individual decision trees, making them more reliable for practical applications.

- **Versatility**: Random forests can handle both numerical and categorical data and perform well even when data has missing values or is not scaled/normalized. They are also capable of dealing with unbalanced datasets effectively.

Limitations

Among the limitations of the random forest classifier, the top three that often pose significant challenges are as follows:

- **Model complexity and interpretability**: With potentially hundreds of trees, random forests create more complex models than a single decision tree, leading to challenges in model interpretation and visualization

- **Computationally intensive**: Training a large number of trees with many features can be computationally demanding, requiring more memory and processing power, and increasing the training time

- **Performance in high-dimensional spaces**: While random forests generally perform well, their effectiveness can diminish in cases with a very high number of features and a relatively small number of samples due to the curse of dimensionality

Despite these limitations, random forest classifiers are highly valued for their robust performance across a wide range of applications, making them a popular choice among data scientists for tackling complex classification problems.

Gradient-boosted tree classifier

Gradient-boosted tree (**GBT**) classifier is a sophisticated machine learning technique that belongs to the ensemble family, specifically under boosting methods. It builds on the principle of boosting, where multiple weak learners (typically decision trees) are combined sequentially to create a strong predictive model. Each tree in the sequence is designed to correct the errors of its predecessor, focusing on the most challenging cases that the previous trees struggled with. This process involves training trees using the gradient of the loss function, which is why it's termed **gradient boosting**.

The core idea behind GBT is to construct new models that predict the residuals or errors of prior models and then combine these predictions through a weighted sum to make the final prediction more accurate. The "gradient" part comes from the use of gradient descent on the loss function to minimize prediction errors. Trees are added one at a time, and existing trees in the model are not changed. After each tree is added, the data weights are updated to focus more on instances that are difficult to predict. This iterative refinement makes the GBT approach highly effective, albeit computationally intensive.

Imagine you're coaching a relay race team with runners of varying skill levels. The team's goal is to win the race, but after the first practice run, it's clear that there are weaknesses: some runners are slower, some have poor baton handoffs, and others tire quickly.

To help them improve, you start by focusing on the most significant issue: the slowest runner. With targeted training, you manage to improve their speed. However, the team is still not winning because now the baton handoffs become the biggest weakness. So, you focus on improving that next. Each round of practice and improvement targets the most pressing issue from the previous run, gradually enhancing the overall performance of the team.

This iterative process of identifying and improving the weakest link after each practice run is analogous to how a GBT classifier works. The "team" is your model, and the "runners" are decision trees. The first tree (runner) is trained and its mistakes are analyzed. The next tree is then trained with a focus on correcting those mistakes, essentially giving more weight to the instances that were misclassified. This process continues, with each new tree focusing on the errors of the ensemble of previous trees, much like each round of targeted training focuses on the team's current biggest weakness.

Just as your relay team becomes stronger and more likely to win with each iteration of focused improvement, the GBT classifier becomes more accurate with each subsequent tree added as it learns from the mistakes of the previous ones, leading to a robust and highly accurate model. The following figure shows the working mechanism of the GBT classifier:

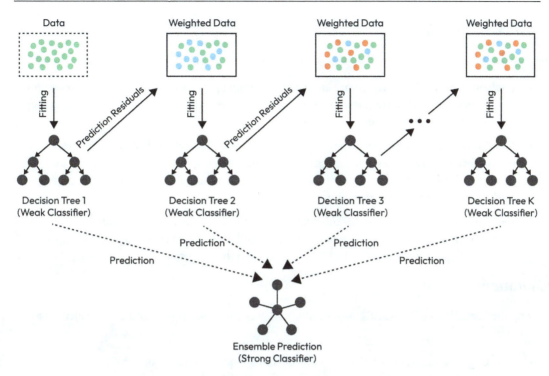

Figure 5.6 – Working mechanism of the GBT classifier

Applications

Here are three notable applications of the GBT classifier:

- **Search engines**: GBT classifiers are integral in web search algorithms as they rank pages based on various features, such as keyword relevance, page views, and links to the page. They help improve the accuracy and relevance of search query results.

- **Financial modeling**: In finance, GBT models are used for credit scoring, predicting the likelihood of default on loans based on the borrower's history and transaction data. They are also employed in algorithmic trading to predict market movements and make trading decisions.

- **Biomedical data analysis**: GBT classifiers are used in genomics and drug discovery to classify patient samples into disease categories or predict drug response based on genetic information, significantly aiding personalized medicine and therapeutic interventions.

Benefits

The top three benefits of the GBT classifier are as follows:

- **High accuracy**: GBT classifiers are known for their high predictive accuracy, even on complex datasets with intricate structures. By focusing on correcting the mistakes of previous trees, they can model complex decision boundaries.

- **Flexibility**: They can handle various types of data (numerical, categorical) and work well with untransformed datasets, including missing values and variables of different scales. Their flexibility also extends to the loss function, which can be tailored to specific problems, enhancing their applicability across diverse domains.

- **Feature importance**: GBT models inherently perform feature selection, giving higher importance to more relevant features. This not only improves model performance but also provides insights into which features are most influential in predicting the outcome.

Limitations

Among the limitations of the GBT classifier, the top three that often pose significant challenges are as follows:

- **Training time and computational cost**: Due to the sequential nature of boosting, GBT models can be time-consuming to train, especially with large datasets and a high number of trees. This can lead to significant computational overhead and energy consumption.

- **Overfitting**: While GBT classifiers are less prone to overfitting than some other models, they can still overfit the training data if the number of trees is too high or the trees are too complex. This necessitates carefully tuning hyperparameters such as tree depth and learning rate.

- **Interpretability**: As with most ensemble methods, the complexity of GBT models makes them harder to interpret than simpler models, such as decision trees. The aggregated nature of the predictions from hundreds or thousands of trees can obscure the decision-making process, making it difficult to understand the model's rationale.

GBT classifiers, with their powerful predictive capabilities and adaptability, have become a mainstay in the machine learning toolbox. However, their effectiveness comes with the caveat of increased model complexity and computational demands, highlighting the trade-offs that practitioners must navigate in their application.

Multilayer perceptron classifier

A **multilayer perceptron** (**MLP**) classifier is a class of feedforward **artificial neural networks** (**ANNs**) that consists of at least three layers of nodes: an input layer, one or more hidden layers, and an output layer. Each node, or neuron, in one layer connects with a certain weight to every node in the following layer, making the network fully connected. MLP utilizes a supervised learning technique called backpropagation to train its networks; it involves adjusting the weights of the connections in the network based on the error rate obtained in the previous epoch (that is, iteration).

The key characteristic of MLPs is the presence of one or more hidden layers, which enable the network to model complex non-linear relationships between the input and output variables. Neurons in these layers apply activation functions (such as sigmoid, tanh, or ReLU) to the weighted inputs they receive, introducing non-linear properties to the network's functioning. This is crucial for learning and modeling complex patterns in the data.

Imagine you're at a large, multi-level amusement park with your friends, trying to decide which rides to go on. The park is divided into different zones (layers), each with its own set of rides (neurons). You start at the entrance (input layer), where you and your friends discuss what you're looking for in a ride: excitement, speed, height, and so on. These preferences are your input features.

As you move into the first zone (the first hidden layer), you're greeted by a group of guides (neurons). Each guide specializes in a certain combination of your desired features: one might know the best rides for both speed and excitement, another might focus on height and speed, and so on. Based on your preferences, each guide gives you a weighted recommendation on where to go next.

Taking this advice, you move to the next zone (the next hidden layer), where a new set of guides awaits. These guides take the recommendations from the first set and combine them, further refining the advice. This process repeats through several zones (hidden layers), with each set of guides integrating the information passed along from the previous ones, progressively honing in on the perfect ride for you.

Finally, you reach the last zone (the output layer), where the ultimate guide (output neuron) takes all the refined advice and points you to the ride that best matches your initial preferences. This final recommendation is the classification outcome of the network.

In this analogy, the MLP classifier works like navigating the amusement park, with each layer of guides (neurons) processing and passing on information, refining the decision at each step based on the input features (your ride preferences) until it reaches the most suitable outcome (the best ride for you and your friends).

Applications

Here are three notable applications of the MLP classifier:

- **Image recognition and classification**: MLPs are widely used in computer vision for tasks such as identifying objects within images, facial recognition, and character recognition in handwriting. Their ability to process and learn from pixel patterns makes them highly effective in these areas.

- **Speech recognition**: In the domain of natural language processing, MLPs contribute significantly to converting speech into text, understanding spoken commands, and even identifying the speaker's identity or emotional state by analyzing the spectral properties of the speech signal.

- **Financial forecasting**: MLPs are employed in predicting stock prices, market trends, and credit risk assessment. They can analyze vast amounts of historical financial data to forecast future financial conditions, aiding in investment decisions and risk management.

Benefits

The top three benefits of the MLP classifier are as follows:

- **Modeling complex relationships**: The architecture of MLPs, with multiple layers and non-linear activation functions, allows them to learn and model highly complex relationships in the data, capturing subtleties that simpler models might miss.

- **Versatility and adaptability**: MLPs can handle a wide range of data types and can be applied to various tasks, from regression to classification, making them highly versatile. They can also be tailored to specific problems by adjusting their architecture, activation functions, and training algorithms.

- **Generalization ability**: When properly trained and regularized, MLPs can generalize well to unseen data, making robust predictions beyond the scope of their training dataset. This is crucial for practical applications where the model encounters new instances in the real world.

Limitations

Among the limitations of the MLP classifier, the top three that often pose significant challenges are as follows:

- **Vulnerability to overfitting**: Due to their complexity and the large number of parameters, MLPs are prone to overfitting, especially when trained on limited or noisy data. Regularization techniques such as dropout and early stopping are often necessary to mitigate this issue.

- **Opaque decision-making process**: The "black box" nature of MLPs makes it challenging to interpret how they make decisions, which can be a significant drawback in applications requiring transparency and explainability, such as in healthcare or legal decisions.

- **Computational intensity**: Training MLPs, particularly deep networks with many layers, can be computationally intensive and time-consuming, requiring substantial hardware resources (such as GPUs) and efficient training algorithms to achieve convergence within a reasonable time frame.

In summary, MLP classifiers, with their deep and intricate architectures, offer powerful tools for capturing and modeling complex patterns in data across various applications. However, their effectiveness is counterbalanced by challenges related to overfitting, interpretability, and computational demands, necessitating careful design, regularization, and resource allocation to fully leverage their potential.

Linear SVM

A linear SVM is a powerful supervised machine learning algorithm that's primarily used for classification tasks, and under certain conditions, for regression as well. It operates on the principle of finding the hyperplane that best separates the classes in the feature space. In the context of classification, the SVM aims to find the hyperplane with the maximum margin, which is defined as the maximum distance between the hyperplane and the nearest data point from either class. This approach helps to ensure that the classifier is not only accurate on the training data but also generalizes well to unseen data.

Imagine that you and your friends are in a large park, and you've brought a variety of balls to play with: footballs and basketballs. You want to organize a game but to make it fair and fun, you decide to separate the footballs and basketballs into two distinct groups. You lay a long rope on the ground, trying to create a line that divides the two types of balls. The goal is to position this rope so that all footballs are on one side and all basketballs are on the other, maximizing the space between the balls and the rope to avoid any mix-ups when choosing a ball for the game.

This rope is like the decision boundary in a linear SVM. Just as you adjust the rope to maximize the margin between the balls and the rope, the linear SVM algorithm seeks to find the hyperplane (in higher-dimensional space, the "rope" becomes a "plane" or "surface") that best separates the different categories in the data (footballs and basketballs in this case) with the maximum margin. The balls closest to the rope, which directly influence its placement, are analogous to the "support vectors" in the SVM, which are the critical elements of the dataset that define the decision boundary. Just as the game becomes clearer and fairer when the balls are well-separated, a classification problem becomes more robust and accurate when the decision boundary separates the different classes with a wide margin.

The following figure shows support vectors that help to define the classification boundary:

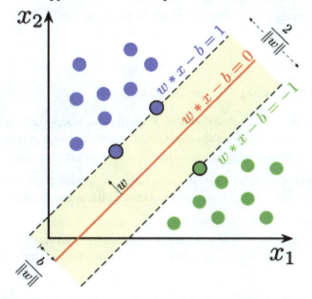

Figure 5.7 – Working mechanism of the linear SVM classifier

The key mechanics of the linear SVM classifier are as follows:

- **Hyperplane**: In a two-dimensional space, the hyperplane is simply a line, but in higher-dimensional spaces, it's a plane or a surface that separates the data points into different classes. The best hyperplane is the one that leaves the widest margin from the nearest points of any class, known as support vectors.

- **Support vectors**: These are the data points nearest to the hyperplane and are critical to defining the hyperplane's position and orientation in the feature space. The algorithm focuses on these points as they are the most difficult to classify and provide the most information about the decision boundary.

- **Margin maximization**: The linear SVM algorithm seeks to maximize the margin between the classes and the hyperplane, leading to a more robust classifier. This is achieved through an optimization process that adjusts the weights and biases to maximize this margin.

Applications

Here are three notable applications of the linear SVM classifier:

- **Text and hypertext categorization**: SVMs are highly effective in classifying texts and hypertexts, making them suitable for applications such as spam detection in emails, topic categorization of news articles, and sentiment analysis in social media content.

- **Image classification**: Linear SVMs, often in conjunction with other feature extraction methods, are used for categorizing images in computer vision tasks, such as face detection in security systems and object recognition in automated systems.

- **Bioinformatics**: In the field of bioinformatics, SVMs are applied to classify proteins, patients based on genetic profiles, and other biological problems where the data can be linearly separated.

Benefits

The top three benefits of the linear SVM classifier are as follows:

- **Effectiveness in high-dimensional spaces**: Even in cases where the number of dimensions exceeds the number of samples, SVMs are known to perform well, making them particularly suitable for text classification and bioinformatics, where this is often the case.

- **Memory efficiency**: Due to their reliance on a subset of training points (the support vectors), SVMs are memory-efficient as they don't need to store the entire dataset.

- **Versatility**: The ability to use different kernel functions makes SVMs highly versatile. Although we're focusing on the linear SVM classifier here, by applying an appropriate kernel function, SVMs can solve not only linearly separable problems but also complex non-linear ones.

Limitations

Among the limitations of the linear SVM classifier, the top three that often pose significant challenges are as follows:

- **Poor performance on overly complex data**: When data is heavily overlapped or not linearly separable, a linear SVM might struggle to find an optimal hyperplane, leading to subpar performance.

- **Selecting a kernel**: Kernels are a way to represent data samples flexibly so that we can compare them in a complex space. The linear SVM classifier uses a linear kernel, but choosing the right kernel (for non-linear SVMs) and tuning its parameters can be complex and is crucial for the algorithm's performance.

- **Scalability**: SVMs, in general, can become computationally intensive and less efficient as the dataset size increases, especially in terms of training time, making them less suitable for very large datasets.

In summary, the linear SVM classifier stands out for its robustness and efficiency in handling high-dimensional data and binary classification tasks, offering precise and reliable predictive modeling across various domains, from text categorization to bioinformatics. However, its effectiveness can be constrained by the linear separability of the data, the challenges in kernel selection for non-linear problems, and scalability issues with large datasets.

The One-vs-Rest classifier (also known as One-vs-All)

The **One-vs-Rest (OvR)** classifier, also known as **One-vs-All**, is a strategy for multi-class classification that involves training a single classifier per class by using the samples of that class as positive samples and all other samples as negatives. This approach is widely used to extend binary classifiers, which are inherently designed to distinguish between two classes, to handle multi-class problems effectively.

In an OvR scheme, if there are N classes, N separate binary classifiers are trained. For example, in a scenario with three classes, A, B, and C, three classifiers would be trained: one to distinguish A from not A (B and C), another to distinguish B from not B (A and C), and a third to distinguish C from not C (A and B). For a given test instance, all N classifiers are applied, and the one that classifies the instance with the highest confidence, typically measured as a probability or distance from the decision boundary, is chosen as the predicted class.

Imagine you're organizing a big talent show with a variety of acts: singers, dancers, magicians, and comedians. To manage the event smoothly, you decide to create separate queues for each type of performer so that they can be judged by a panel specialized in their respective fields. However, you only have enough resources to set up one judging panel at a time.

So, you start with the singers: you set up a singers' queue and a not-singers' queue, which includes all the other acts. The singers perform, and the judges evaluate them. Once done, you switch the panel to judge dancers. Now, the dancers' queue is formed, and everyone else, including the previously judged singers and the yet-to-be-judged magicians and comedians, goes into the not-dancers' queue. This process repeats until each category of performers has been judged by their specialized panel.

This approach mirrors the OvR classification strategy used in machine learning. In a multi-class classification problem, the OvR method involves training a separate binary classifier for each class to distinguish that class from all other classes. For instance, one classifier might be trained to identify singers by distinguishing them from non-singers (dancers, magicians, comedians), another to identify dancers from non-dancers, and so on. Each classifier's task is to focus on separating its target class from the rest, much like setting up a dedicated judging panel for each type of talent show act. This way, despite having multiple categories, you efficiently manage to evaluate each one using a series of binary, or two-group, decisions:

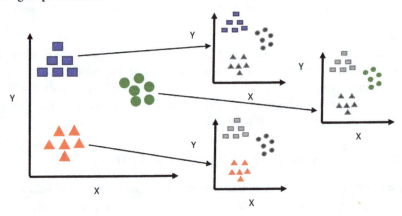

Figure 5.8 – Working mechanism of the OvR classifier

Applications

Here are three notable applications of the OvR classifier:

- **Text classification**: OvR is commonly used in classifying documents or emails into multiple categories based on their content. Each category is treated as a separate class, and OvR classifiers are trained to identify documents belonging to each category against all others.

- **Image recognition**: In scenarios where each image needs to be classified into one of multiple categories (for example, animals, vehicles, and landscapes), OvR can be applied to train individual classifiers to recognize each category against all other images.

- **Medical diagnosis**: OvR classifiers can be employed to diagnose multiple types of diseases from medical images or patient data, where each disease represents a distinct class, and the classifier for each disease distinguishes it from healthy conditions and other diseases.

Benefits

The top three benefits of the OvR classifier are as follows:

- **Simplicity and scalability**: The OvR strategy is straightforward to implement and understand, making it an accessible approach for extending binary classification algorithms to multi-class problems. It's particularly effective for dealing with a large number of classes as it scales linearly with the number of classes.

- **Flexibility**: OvR can be applied to virtually any binary classification algorithm, allowing practitioners to leverage a wide array of existing algorithms and tailor their approach to the specific characteristics of their dataset and problem.

- **Parallelization**: Since each classifier is trained independently of the others, OvR lends itself well to parallelization, significantly reducing training time when computational resources allow multiple classifiers to be trained simultaneously.

Limitations

Among the limitations of the OvR classifier, the top three that often pose significant challenges are as follows:

- **Class imbalance**: In the OvR approach, each classifier is trained on a dataset that's imbalanced by design, with one class against all others. This can lead to classifiers that are biased toward the majority "negative" class, particularly in cases where there's a significant disparity in the size of the classes.

- **Increased computational burden**: Training multiple classifiers can be computationally expensive, especially as the number of classes increases. This includes not only the computational cost during training but also the storage requirements for multiple models and the inference time when making predictions.

- **Inter-class correlation**: OvR classifiers treat each class in isolation, which can be a drawback when there are correlations or dependencies between classes. This isolation means that the classifiers may not leverage potentially useful information about the relationships between classes, potentially leading to suboptimal performance compared to methods that consider all classes simultaneously.

Despite these limitations, the OvR strategy remains a popular and effective approach for multi-class classification, offering a balance between simplicity, flexibility, and performance that makes it suitable for a wide range of applications.

Naive Bayes

Naive Bayes classifiers are a family of simple "probabilistic classifiers" based on Bayes' theorem with strong independence assumptions between the features. They are highly efficient and particularly well-suited for high-dimensional datasets. Despite their simplicity, Naive Bayes models have shown remarkable performance in various tasks.

Naive Bayes classifiers assume that the value of a particular feature is independent of any other feature value, given the class variable. This assumption is called **class conditional independence**. For example, a fruit is considered to be an apple if it is red, round, and about 2 to 3 inches in diameter. A Naive Bayes classifier considers these features to contribute independently to the probability that this fruit is an apple, regardless of any possible correlations between the color, roundness, and diameter features.

The model calculates the posterior probability of each class, based on the input features, using Bayes' theorem. In practice, the denominator remains constant for all classes, so it's only necessary to focus on the numerator, which involves the prior probability of the class and the likelihood of the data given the class. The class with the highest posterior probability is considered the predicted class.

Imagine that you're a detective trying to solve a mystery based on a set of clues. Each clue, on its own, might point to various suspects, but the real skill lies in combining these clues to identify the most likely culprit.

The following figure shows the working of the Naive Bayes classifier in terms of classifying several shapes as square, triangle, and circle:

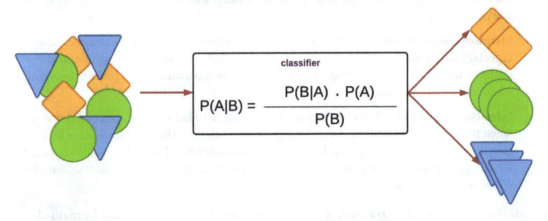

Figure 5.9 – Working mechanism of the Naive Bayes classifier

Consider a case where you're trying to determine whether a person is the thief of a stolen painting. You have various clues, including that the thief has paint on their shoes, they left a glove at the scene, and a witness saw a tall person sneaking around. You've seen many cases, so you have a good idea of how common each of these clues is among thieves and innocent people.

You start by considering how likely it is for someone to have paint on their shoes. Maybe it's common among painters but rare for others. Then, you think about the glove. It's winter, so lots of people could be wearing gloves, but maybe this type of glove is unique. Lastly, you consider height. Being tall might not be very uncommon.

In your detective work, you assume that each clue independently contributes to the likelihood of someone being the thief, without considering how the clues might be related to each other. For example, you don't assume that just because someone is tall, they're more likely to have paint on their shoes. This is the "naive" part of Naive Bayes.

Like piecing together clues, the Naive Bayes classifier calculates the probability of someone being the thief (or a certain category in a classification problem) based on the independent contributions of each clue (or feature). It combines these probabilities to find the most likely category, just as you would combine clues to find the most likely suspect. Despite its simplicity and the "naive" assumption of independence, just like a skilled detective, Naive Bayes can be surprisingly effective in solving the mystery – or classification problem – at hand.

Applications

Here are three notable applications of the Naive Bayes classifier:

- **Spam detection**: One of the classic applications of Naive Bayes is email spam filtering. The algorithm assesses the presence or absence of certain words that are indicative of spam or non-spam emails and categorizes them accordingly.

- **Document classification**: Naive Bayes classifiers are widely used in text categorization, determining whether a document belongs to one or more categories (such as sports, politics, or entertainment) based on the frequency of specific words or phrases.

- **Sentiment analysis**: They are also employed in analyzing the sentiment of text data, such as reviews or social media posts, to classify them as positive, negative, or neutral based on the presence of certain keywords.

Benefits

The top three benefits of the Naive Bayes classifier are as follows:

- **Simplicity and efficiency**: Naive Bayes classifiers are straightforward to implement and require a small amount of training data to estimate the necessary parameters, making them highly efficient, especially for large datasets

- **Performance**: Despite the simplicity and the naive assumption, Naive Bayes classifiers can outperform more sophisticated algorithms, particularly in text classification tasks and when the independence assumption holds reasonably true

- **Handle high-dimensional data**: They are particularly adept at handling high-dimensional data efficiently, making them ideal for text classification where the feature space can be extremely large (for example, bag-of-words models)

Limitations

Among the limitations of the Naive Bayes classifier, the top three that often pose significant challenges are as follows:

- **Naive assumption**: The assumption of feature independence is rarely true in real-world applications, which can limit the classifier's accuracy. In complex problems where feature relationships are important, this assumption can lead to suboptimal performance.

- **Data scarcity for probability estimation**: Naive Bayes classifiers rely on the frequency of the features' values to calculate probabilities. When a frequency is zero because the training data doesn't contain a particular feature-class combination, it can lead to incorrect predictions. This is often mitigated by techniques such as Laplace smoothing.

- **Difficulty with continuous features**: Naive Bayes inherently works with frequency counts. When dealing with continuous data, assumptions have to be made to fit the model (for example, assuming a Gaussian distribution), which may not always align well with the data's actual distribution.

Despite these limitations, the Naive Bayes classifier remains a popular choice due to its simplicity, efficiency, and surprisingly good performance in various tasks, particularly in text processing and analysis.

Factorization machines classifier

Factorization machines (**FMs**) are versatile machine learning models that are designed to capture interactions between variables in high-dimensional sparse datasets efficiently. They are particularly powerful in scenarios where the data involves categorical variables that have been converted into a large number of binary features through techniques such as one-hot encoding. FMs combine the advantages of SVMs with factorization models, making them adept at dealing with data that's common in recommendation systems, computational advertising, and natural language processing.

FMs model interactions between variables using factorized parameters. This means that instead of having a separate parameter for each interaction term (which would be computationally infeasible in high-dimensional spaces), FMs learn a low-dimensional vector for each variable. The interaction between any two variables is then modeled as the dot product of their corresponding vectors. This approach significantly reduces the number of parameters to be learned, making the model scalable to very high-dimensional datasets while still being able to capture complex interactions between variables.

Imagine you're planning a giant potluck dinner, aiming to personalize dish recommendations for each guest based on their known food preferences, previous potluck dishes they enjoyed, and how similar guests have rated dishes. This complex scenario, with multiple interacting factors, can be likened to the workings of the FMs classifier.

Consider each guest's preference as a unique feature, and each dish's attributes (such as cuisine type, key ingredients, and spiciness level) as another set of features. A simple approach might be to match guests with dishes based on direct overlaps in preferences and dish attributes (for example, someone likes spicy food, so recommend a spicy dish). However, this ignores more subtle interactions – maybe a guest likes spicy food but only in Mexican cuisine, or they particularly enjoy dishes that other guests with similar tastes have liked.

FMs shine in this context by not just considering these direct relationships but also learning the hidden patterns in how these features interact. For instance, the FM classifier might uncover that guests who like spicy Mexican dishes often also enjoy certain Indian curries, even if they haven't explicitly stated a preference for Indian food. It does this by factorizing the massive, complex matrix of all possible guest-dish interactions into lower-dimensional representations, capturing the essence of these interactions in a more manageable form.

In the potluck analogy, using the FMs classifier is like having a super-smart organizer who remembers every guest's past dish ratings, understands each dish in detail, and knows how guests' preferences subtly relate to each other. This organizer then makes nuanced dish recommendations for each guest, considering not just their direct preferences but also the deeper patterns of what they and similar guests have enjoyed in the past.

Just as this organizer would elevate the potluck experience by personalizing it to an unprecedented degree, FMs provide a powerful tool in machine learning for tackling problems with rich, interrelated features, offering highly personalized and accurate classifications.

The following figure shows the working mechanism of the FMs classifier. In this example, a matrix of user ratings on items is taken as input to generate latent factor embeddings. These embeddings are used to predict ratings.

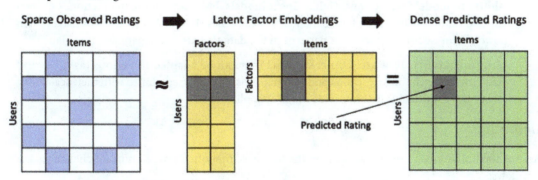

Figure 5.10 – Working mechanism of the FMs classifier

Applications

Here are three notable applications of the FMs classifier:

- **Recommendation systems**: FMs are widely used in recommendation systems, where they excel at predicting user preferences based on past interactions. They can efficiently handle the massive, sparse datasets typical of user-item matrices in these systems, capturing the subtle interactions between users and items.

- **Click-through rate (CTR) prediction**: In computational advertising, FMs are employed to predict the probability that a given advert will be clicked by a user. They can take into account user demographics, advert features, and contextual information, modeling the interactions between these features to improve prediction accuracy.

- **Sentiment analysis and text mining**: By representing text as high-dimensional sparse vectors (for example, using TF-IDF), FMs can be applied to sentiment analysis and other text mining tasks, capturing the interactions between different words or phrases to better understand the context and sentiment of the text.

Benefits

The top three benefits of the FMs classifier are as follows:

- **Efficiency in high-dimensional spaces**: FMs are specifically designed to work efficiently in situations where the data is high-dimensional and sparse. They can handle millions of features without a significant increase in computational requirements thanks to their factorized representation of interactions.

- **Ability to model interactions**: Unlike linear models, FMs can capture interactions between any two features without needing to manually specify these interaction terms. This makes them powerful for uncovering complex patterns in the data that are not apparent at first glance.

- **Generalizability**: FMs are a general model that can be adapted to various tasks beyond classification, including regression and ranking. This versatility, combined with their scalability, makes them a valuable tool in a wide range of applications.

Limitations

Among the limitations of the FMs classifier, the top three that often pose significant challenges are as follows:

- **Model complexity and overfitting**: While FMs are efficient in terms of the number of parameters, the model itself can become complex, especially with a large number of latent factors. This complexity can lead to overfitting, particularly when the amount of training data is not sufficiently large to support the model.

- **The computational cost for training**: Despite their efficiency in handling high-dimensional data, the training process for FMs, especially with large datasets and a high number of latent factors, can be computationally intensive and time-consuming compared to simpler models.

- **Hyperparameter sensitivity**: The performance of FMs can be highly sensitive to the choice of hyperparameters, such as the number of latent factors and regularization terms. Finding the optimal set of hyperparameters often requires extensive search and cross-validation, which can be a time-consuming process.

In conclusion, FMs offer a powerful and flexible approach for dealing with high-dimensional sparse datasets, particularly where interactions between features play a crucial role. Their applications in recommendation systems, CTR prediction, and text mining showcase their versatility and effectiveness. However, their complexity and computational demands highlight the need for careful model tuning and consideration of the available computational resources.

Evaluating the model's performance

Evaluating the performance of classification models is crucial to understanding their effectiveness and suitability for specific tasks. Various metrics have been developed to capture different aspects of model performance, each with its unique insights and applications.

Binary classification

In Apache Spark, for binary classification tasks, the MLlib library provides several key evaluation metrics that can be used to assess the performance of models. These metrics help in understanding various aspects of models, including their accuracy, precision, recall, and the trade-offs between them. The primary evaluation metrics that are available in Spark for binary classification include the following:

- **Accuracy**: While not always the most informative metric for binary classification, especially in imbalanced datasets, accuracy measures the proportion of correct predictions (both true positives and true negatives) among all predictions made. It is calculated as $((TP + TN) / (TP + FP + FN + TN))$.

- **Precision**: Precision (or positive predictive value) measures the proportion of true positive predictions in all positive predictions made by the model. It indicates the model's accuracy in identifying positive instances. It is calculated as $(TP / (TP + FP))$.

- **Recall (sensitivity or true positive rate (TPR)**: Recall measures the proportion of true positive predictions out of all actual positive instances. It assesses the model's ability to capture or identify positive instances. It is calculated as $(TP / (TP + FN))$.

- **F1 score**: The F1 score is the harmonic mean of precision and recall. It is particularly useful when you need to balance precision and recall.

- **Area under receiver operating characteristic (ROC) curve (AUC-ROC):** AUC-ROC is a widely used metric that measures the model's ability to discriminate between positive and negative classes. The ROC curve plots the **TPR** against the **false positive rate (FPR)** at various threshold settings. The AUC represents the probability that a randomly chosen positive instance is ranked higher than a randomly chosen negative instance:

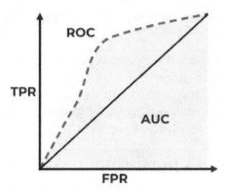

Figure 5.11 – ROC curve and AUC

- **Area under precision-recall curve (AUC-PR):** The precision-recall curve is particularly informative for imbalanced datasets, where the positive class is much rarer than the negative class. AUC-PR summarizes the trade-off between precision and recall for different thresholds, with a focus on the performance of the minority class.

To compute these metrics in Spark, you can use BinaryClassificationEvaluator for AUC-ROC and MulticlassClassificationEvaluator for precision, recall, F1 score, and accuracy. While BinaryClassificationEvaluator is specifically designed for binary classification tasks, MulticlassClassificationEvaluator can be used for both binary and multiclass classification problems, and you would specify the metric name you're interested in calculating.

It's important to choose the evaluation metric(s) that best align with the objectives and context of your specific problem, especially while considering factors such as class imbalance or the cost of different types of errors (false positives versus false negatives). Spark provides the flexibility to calculate and analyze these metrics, aiding in thoroughly evaluating and tuning binary classification models.

The following figure shows the basics of model evaluation. In this example, when both the prediction and ground truth are sunny, then it is a true positive:

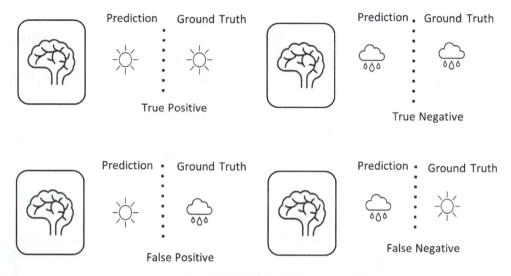

Figure 5.12 – Model evaluation

Multiclass classification

In Apache Spark, for multiclass classification tasks, the MLlib library provides several key evaluation metrics that can be used to assess the performance of models.

Apache Spark's MLlib offers a variety of evaluation metrics for classification tasks, all of which can help assess the performance of models in different ways:

- **Confusion matrix**: A table showing the actual versus predicted classifications. It includes **true positives (TP)**, **false positives (FP)**, **true negatives (TN)**, and **false negatives (FN)**, providing a comprehensive overview of the model's performance:

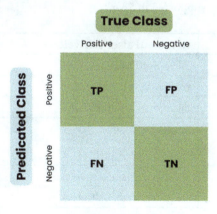

Figure 5.13 – Confusion matrix

- **Log loss**: Though not directly provided by Spark's MLlib, it measures the performance of a classification model where the prediction is a probability between 0 and 1. It penalizes incorrect classifications based on the confidence of the prediction.

- **Weighted F-measure**: A variant of F-measure that accounts for class imbalance by weighting the F1 score of each class by the number of true instances, providing a metric that's more reflective of real-world performance.

- **Weighted FPR**: An average of FPRs across all classes, weighted by the number of true instances in each class. It accounts for class imbalance in assessing how often negative instances are incorrectly classified as positive.

- **Weighted precision**: Similar to precision but weighted by the number of true instances for each class, thus accounting for class imbalance in assessing the accuracy of positive predictions.

- **Weighted recall (weighted TPR)**: The recall for each class, weighted by the number of true instances in each class. It provides an overall measure of the model's ability to correctly identify positive instances while accounting for class imbalance.

- **Hamming loss**: Hamming loss is a measure for multi-label classification models where the prediction error is a fraction of the labels that are incorrectly predicted – that is, the fraction of the wrong labels to the total number of labels.

To compute these metrics in Spark, you would typically use MulticlassClassificationEvaluator, specifying the metric name you're interested in (for example, `f1`, `precision`, `recall`, or `accuracy`). Spark allows you to assess these metrics either for the entire model or for specific classes, giving you a comprehensive view of your model's performance across multiple classes.

When evaluating a multiclass classification model in Spark, it's important to select metrics that align with your specific problem's needs, especially while considering factors such as the relative importance of classes or the cost associated with different types of misclassifications.

Here's a list of evaluation metrics supported in Spark:

Accuracy	weightedFalsePositiveRate	PrecisionByLabel
WeightedPrecision	WeightedFMeasure	RecallByLabel
WeightedRecall	truePositiveRateByLabel	fMeasureByLabel
weightedTruePositiveRate	falsePositiveRateByLabel	LogLoss
F1	hammingLoss	confusionMatrix
	falsePositiveRate	truePositiveRate
weightedFMeasure	weightedFalsePositiveRate	weightedPrecision
weightedRecall	weightedTruePositiveRate	

Table 5.1 – Evaluation metrics

Algorithm-specific considerations

Please note that model performance evaluation metrics cannot be used in all the algorithms. Here are some considerations for selecting the metrics based on the algorithm:

- **Decision tree, random forest, and GBT**: Given these models' capacity for both binary and multiclass problems, you can use the respective binary or multiclass metrics based on your specific use case.

- **MLP classifier**: Primarily used for multiclass classification, so multiclass metrics are applicable.

- **OvR**: Depending on whether it's set up for a binary or multiclass problem, you can use either binary classification metrics or multiclass metrics.

- **FMs classifier**: While Spark's MLlib does not natively include FMs, if implemented or used through an extension, you can evaluate it using binary or multiclass metrics depending on your specific application.

Selection tips

The following are some tips you can use to select appropriate evaluation metrics:

- **Consider your problem type**: Choose binary or multiclass metrics based on the nature of your classification problem.

- **Imbalanced datasets**: In cases of imbalanced data, metrics such as AUC-PR, weighted F1 score, or precision and recall for the minority class become more critical.

- **Domain-specific requirements**: Align metric selection with domain needs. For example, if false negatives are more costly than false positives, prioritize recall.

Selecting the evaluation metrics

Selecting the right evaluation metrics for classification tasks involves considering several key factors related to the specific problem you're solving, the nature of your dataset, and the business or research objectives. Here's a guide to help you choose the most appropriate metrics.

Understand your business objectives

The choice of metric should align with your project's ultimate goals. Different applications might prioritize different types of errors:

- In a spam detection system, false positives (marking important emails as spam) might be more problematic than false negatives (letting some spam through)

- In medical diagnosis, missing a positive case (false negatives) could be far more critical than false alarms (false positives)

Consider the nature of your data

It is very important to understand the data distribution and the context before selecting the appropriate metrics:

- **Class distribution**: If your dataset is imbalanced (that is, one class significantly outnumbers another), accuracy might not be a reliable metric. In such cases, precision, recall, F1 score, or AUC might provide more insight.

- **Multi-class versus binary classification**: For multi-class problems, you might need to consider metrics that can handle multiple classes effectively, such as macro-averaged or micro-averaged F1 scores, or confusion matrices to visualize performance across all classes.

Type of error to prioritize

Decide which type of error is more critical to minimize in your application:

- **False Positives**: If minimizing false alarms is crucial, you should prioritize precision

- **False Negatives**: If it's more important to capture all positive cases, focus on recall

You might need a balance between these, in which case the F1 score could be ideal as it harmonizes precision and recall.

Model comparison and selection

If you're comparing multiple models or need a single summary metric, consider the following:

- AUC-ROC is useful for comparing different models, especially in imbalanced datasets, as it evaluates model performance across all classification thresholds

- Log loss provides a way to compare models based on the probabilities they output, penalizing confident wrong predictions more harshly

Real-world constraints

Consider operational constraints such as the following:

- **Interpretability**: In some applications, understanding why a model made a certain prediction is crucial. Metrics derived from confusion matrices (such as precision and recall) are more interpretable.

- **Computational resources**: Some metrics might be more computationally intensive to calculate, especially for large datasets or in real-time applications.

Statistical considerations

In scenarios where statistical properties of the metric are important (for example, for significant testing or confidence intervals), metrics such as log loss or MCC might be preferred due to their well-understood statistical properties.

Regulatory and ethical considerations

In regulated industries (such as finance or healthcare) or when ethical considerations are paramount (such as fairness or bias toward certain groups), it's important to choose metrics that can capture and highlight these aspects.

Implementation and validation

After selecting your metrics, you must implement them correctly and validate their reliability:

- Use libraries and tools that are well-tested and widely accepted in the community

- Cross-validate your results to ensure that the chosen metrics provide consistent insights across different subsets of your data

This concludes our learning about evaluation metrics. In the next section, we'll look at various methods we can use to improve the model's performance.

Improving the model's performance

Improving the performance of a classification model involves applying a combination of techniques at different stages of the model development process. Here are some strategies you can use to enhance your model's accuracy and effectiveness:

- **Data quality and quantity**:

  - **Increase data size**: More data can help the model learn better and generalize well to unseen data. Consider augmenting your dataset if possible.

  - **Data cleaning**: Remove outliers, handle missing values, and correct errors in your dataset to improve model accuracy.

  - **Feature engineering**: Create new features from existing ones through domain knowledge. Well-designed features can significantly improve model performance.

- **Feature selection and dimensionality reduction**:

  - **Feature selection**: Identify and retain the most informative features while removing irrelevant or redundant ones to reduce overfitting and improve model performance

  - **Dimensionality reduction**: Techniques such as **principal component analysis (PCA)** can reduce the feature space, potentially improving model efficiency and performance by removing noise

- **Choose the right model**:

  - **Experiment with different models**: Different models have different strengths and weaknesses. Experiment with various algorithms (such as decision trees, SVMs, and ensemble methods) to find the best fit for your data.

- **Hyperparameter tuning**:

  - **Grid search and random search**: These are systematic ways to search through multiple combinations of parameter values, finding the optimal settings for your model

  - **Cross-validation**: Use cross-validation to ensure that your model's performance is robust across different subsets of your dataset

- **Handle imbalanced data**:

  - **Resampling techniques**: Over-sample the minority class or under-sample the majority class to balance the dataset and improve model performance on minority classes

  - **Cost-sensitive learning**: Modify the algorithm to penalize misclassifications of the minority class more than the majority class

  - **Use appropriate metrics**: Choose evaluation metrics that provide insights into model performance on imbalanced datasets, such as F1 score, precision-recall AUC, or Matthews correlation coefficient

- **Advanced feature engineering**:

  - **Interaction terms**: Consider adding features that capture interactions between other features if you suspect such interactions could be predictive

  - **Polynomial features**: Use polynomial features to model non-linear relationships

- **Model regularization**:

  - **Apply regularization techniques**: Techniques such as L1 (lasso) and L2 (ridge) regularization can prevent overfitting by penalizing large coefficients in the model

- **Data transformation**:

  - **Feature scaling**: Standardize or normalize your features so that they're on the same scale. This is particularly important for models sensitive to the scale of the data, such as SVMs or KNN.

  - **Data encoding**: Properly encode categorical variables using techniques such as one-hot encoding, target encoding, or embeddings, especially for algorithms that require numerical input.

- **Model updating**:

  - **Incremental learning**: Continuously update your model with new data so that it can adapt to changes over time

  - **Feedback loops**: Incorporate model predictions back into training to correct errors and adapt to new patterns

- **Domain-specific techniques**:

  - **Utilize domain knowledge**: Incorporating expert knowledge can guide feature engineering, model selection, and interpretation of the results

  - **Custom loss functions**: Design loss functions that specifically address the business problem's nuances

- **Ensemble techniques**: Ensemble methods reduce bias and variance, and also improve generalization. Experiment with these techniques to boost your model's performance:

  - **Bagging (bootstrap aggregating)**:

    - **Random forests**: Combines decision trees by averaging their predictions

    - **Bootstrap sampling**: Creates diverse subsets of the data for each tree

  - **Boosting**:

    - **AdaBoost**: Iteratively trains weak models, emphasizing misclassified samples

    - **Gradient boosting (GBM)**: Builds trees sequentially, correcting errors from the previous tree

    - **XGBoost and LightGBM**: Enhanced versions of GBM with regularization and parallelization

  - **Stacking**: Combines predictions from different models using a meta-model

Improving a classification model is an iterative process that involves experimenting with different strategies, continuously monitoring performance, and adapting to new data and insights. Balancing the trade-offs between model complexity, interpretability, and performance is key to developing effective and reliable classification models.

In the next section, we'll solve the classification problem using a real-world dataset.

Code example

For our example, we'll use the AI4I 2020 Predictive Maintenance Dataset ((2020) UCI Machine Learning Repository (`https://doi.org/10.24432/C5HS5C`). This synthetic dataset is modeled after an existing milling machine and consists of 10,000 data points stored as rows with 14 features in columns:

- UID: unique identifier

- product ID

- type: just the product type L, M, or H

- air temperature [K]

- process temperature [K]

- rotational speed [rpm]

- torque [Nm]

- tool wear [min]: The quality variants, H/M/L, add 5/3/2 minutes of tool wear to the used tool in the process

- The `machine failure` label: This indicates whether the machine has failed in this particular data point if any of the following failure modes are true

The following lines of code imports all the required packages and libraries:

```
from pyspark.sql import SparkSession
spark = SparkSession.builder.appName(
    "ClassificationExample"
).getOrCreate()
```

The following code snippet loads the dataset and prints out the DataFrame and its schema:

```
train_df = spark.read.csv("s3a://test234/machine_failure_data.csv",
    header=True, inferSchema=True)
train_df.show(3, truncate=False)
```

The output is as follows:

```
-------------+------------+-----------------+---------------+
|UDI|Product ID|type|air_temperature_k|process_temperature_k|rotational_speed_rpm|torque_nm|tool_wear_min|machine_failure|tool_wear_failure|heat_
ation_failure|power_failure|overstrain_failure|random_failures|
+---+----------+----+-----------------+---------------------+--------------------+---------+-------------+---------------+-----------------+--------------------+-----------+
-------------+------------+-----------------+---------------+
|1  |M14860    |M   |298.1            |308.6                |1551                |42.8     |0            |0              |0                |0                   |0          |
|0  |          |0   |                 |0                    |                    |
|2  |L47181    |L   |298.2            |308.7                |1408                |46.3     |3            |0              |0                |0                   |0          |
|0  |          |0   |                 |0                    |                    |
|3  |L47182    |L   |298.1            |308.5                |1498                |49.4     |5            |0              |0                |0                   |0          |
|0  |          |0   |                 |0                    |                    |
+---+----------+----+-----------------+---------------------+--------------------+---------+-------------+---------------+-----------------+--------------------+-----------+
-------------+------------+-----------------+---------------+
```

Figure 5.14 – DataFrame output

The following code snippet converts integer columns into floats, removes the null values, and prints the data schema:

```
# Convert all integer columns in train_df to float
from pyspark.sql.types import IntegerType, FloatType
for col in train_df.columns:
    if train_df.schema[col].dataType == IntegerType():
        train_df = train_df.withColumn(col,
            train_df[col].cast(FloatType()))
# remove null values
train_df = train_df.dropna()
train_df.printSchema()
```

Let's see the output:

```
root
 |-- UDI: float (nullable = true)
 |-- Product ID: string (nullable = true)
 |-- type: string (nullable = true)
 |-- air_temperature_k: double (nullable = true)
 |-- process_temperature_k: double (nullable = true)
 |-- rotational_speed_rpm: float (nullable = true)
 |-- torque_nm: double (nullable = true)
 |-- tool_wear_min: float (nullable = true)
 |-- machine_failure: float (nullable = true)
 |-- tool_wear_failure: float (nullable = true)
 |-- heat_dissipation_failure: float (nullable = true)
 |-- power_failure: float (nullable = true)
 |-- overstrain_failure: float (nullable = true)
 |-- random_failures: float (nullable = true)
```

Figure 5.15 – Data schema

The following code snippet performs feature engineering by removing the unwanted columns, replacing the missing values, and converting the categorical column into a numerical one:

```
train_df = train_df.drop("UDI","Product ID",
    "heat_dissipation_failure","power_failure",
    "overstrain_failure","random_failures", "tool_wear_failure")
train_df.show(3,truncate=False)
#replace  the missing values with 0
train_df = train_df.fillna(0)
# convert the type column to index
from pyspark.ml.feature import StringIndexer
indexer = StringIndexer(inputCol="type", outputCol="type_index")
train_df = indexer.fit(train_df).transform(train_df)
train_df.show(3)
```

The following output is generated:

```
+----+-----------------+---------------------+---------------------+---------+-------------+---------------+----------+
|type|air_temperature_k|process_temperature_k|rotational_speed_rpm|torque_nm|tool_wear_min|machine_failure|type_index|
+----+-----------------+---------------------+---------------------+---------+-------------+---------------+----------+
|   M|            298.1|                308.6|              1551.0|     42.8|          0.0|            0.0|       1.0|
|   L|            298.2|                308.7|              1408.0|     46.3|          3.0|            0.0|       0.0|
|   L|            298.1|                308.5|              1498.0|     49.4|          5.0|            0.0|       0.0|
+----+-----------------+---------------------+---------------------+---------+-------------+---------------+----------+
```

Figure 5.16 – DataFrame output

The following code snippet generates the feature through the vector function and normalizes the column:

```
from pyspark.ml.feature import VectorAssembler
assembler = VectorAssembler(inputCols=["air_temperature_k",
        "process_temperature_k", "rotational_speed_rpm",
        "torque_nm", "tool_wear_min","type_index"],
    outputCol="features")
train_df = assembler.transform(train_df)
# apply standard scaler
from pyspark.ml.feature import StandardScaler
scaler = StandardScaler(inputCol="features", outputCol="scaledFeatures")
scalerModel = scaler.fit(train_df)
train_df = scalerModel.transform(train_df)
# show the scaled features
train_df.select("scaledFeatures", "machine_failure").show(3)
```

We'll see the following output:

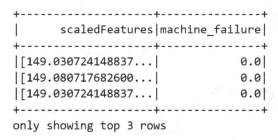

```
+--------------------+---------------+
|      scaledFeatures|machine_failure|
+--------------------+---------------+
|[149.030724148837...|            0.0|
|[149.080717682600...|            0.0|
|[149.030724148837...|            0.0|
+--------------------+---------------+
only showing top 3 rows
```

Figure 5.17 – Scaled features

The following code snippet splits the dataset into train and test tests. It then trains the model using LogisticRegression and evaluates the model's performance.

```
# split the data
train_df, test_df = train_df.randomSplit([0.7, 0.3])
# select the features and label
train_df = train_df.select("scaledFeatures", "machine_failure")
test_df = test_df.select("scaledFeatures", "machine_failure")
```

```
# train the model
from pyspark.ml.classification import LogisticRegression
lr = LogisticRegression(
    featuresCol="scaledFeatures", labelCol="machine_failure")
lrModel = lr.fit(train_df)
# evaluate the model
from pyspark.ml.evaluation import BinaryClassificationEvaluator
evaluator = BinaryClassificationEvaluator(
    labelCol="machine_failure")
evaluator.evaluate(lrModel.transform(test_df))
```

We'll get 0.918577395158542 as output.

This concludes the walk-through of the code example.

Summary

In this chapter, we explored the fundamental concepts and techniques of classification within the realm of supervised learning. Classification stands as a pivotal task in machine learning, allowing data to be classified into predefined classes. This is essential for a wide array of applications, such as email filtering, medical diagnosis, and customer segmentation.

We delved into various classification algorithms, including decision trees, SVM, KNN, logistic regression, Naive Bayes, and FMs. Each algorithm's unique strengths, applications, benefits, and limitations were discussed, providing a comprehensive understanding of their practical use cases.

Through practical case studies and real-world examples, we demonstrated the transformative impact of classification across different sectors, including finance, healthcare, and cybersecurity. We also covered critical aspects of evaluating classifier performance using metrics such as accuracy, precision, recall, F1 score, and the confusion matrix, ensuring rigorous assessment and improvement of models.

Furthermore, we explored advanced techniques for improving model performance, such as data preprocessing, feature engineering, handling data imbalance, hyperparameter tuning, and regularization. Hands-on exercises with code examples in Apache Spark equipped you with the practical skills needed to implement and optimize classification models.

At this point, you should possess a robust understanding of classification and be well-prepared to apply these techniques to address complex problems, marking a significant milestone in your journey through the fascinating world of machine learning.

In the next chapter, we'll learn about clustering.

Part 3:
Unsupervised Learning

In this part, you will dive into the world of unsupervised learning – a powerful branch of machine learning where the model learns from unlabeled data. Unlike supervised learning, where the goal is to predict a target variable, unsupervised learning focuses on uncovering hidden patterns, relationships, and structures within the data. This part is designed to provide you with a thorough understanding of the key concepts, techniques, and algorithms used in unsupervised learning.

The chapters in this part will guide you through clustering techniques, recommendation techniques, and association rule mining. You will explore how these methods are applied in real-life scenarios, allowing you to leverage unsupervised learning to solve complex problems.

This part contains the following chapters:

- *Chapter 6, Building a Clustering System*
- *Chapter 7, Building a Recommendation System*
- *Chapter 8, Mining Frequent Patterns*

6

Building a Clustering System

In the evolving landscape of machine learning, unsupervised learning stands as a beacon of exploration, where algorithms delve into the raw, unlabeled data, seeking patterns, structures, and insights without explicit labels or instructions. This chapter is dedicated to the art and science of clustering, a type of unsupervised learning that groups data points based on their similarities, revealing the underlying structure of the dataset.

Clustering algorithms are the cartographers of data, charting the hidden territories within datasets. They operate under a simple yet profound premise: to group data points together that are alike and separate those that differ significantly. This seemingly straightforward task is rich with complexity and nuance, as the definition of "similarity" can vary widely depending on the context, data type, and desired outcomes.

In this chapter, we explore clustering techniques by examining several algorithms to unhide distinct patterns across data landscapes. We delve into their mechanics, assumptions, strengths, and limitations. Challenges include determining optimal cluster numbers, handling high-dimensional data, and assessing clustering quality. Metrics guide evaluation, while practical applications span market segmentation, anomaly detection, bioinformatics, and social network analysis. Clustering extracts value, informs decisions, and uncovers hidden patterns.

In this chapter, we will cover the following topics:

- Learning about clustering
- Learning clustering algorithms
- Evaluating the model performance
- Improving the model performance

By understanding the principles, challenges, and applications of clustering, you'll be equipped to harness its potential in your own work, pushing the boundaries of what can be discovered in datasets without predefined labels or structures.

Technical requirements

The code files for this chapter are available on GitHub at `https://github.com/PacktPublishing/Apache-Spark-for-Machine-Learning/tree/main/Chapter06`.

Learning about clustering

Clustering is a fundamental technique in unsupervised machine learning and data mining, aimed at grouping a set of objects in such a way that objects in the same group, or cluster, are more like each other than those in other clusters. It's a method of exploratory data analysis used across various fields and applications to discover underlying patterns, categorize data, and summarize datasets by grouping similar items.

Understanding clustering

Unlike supervised learning, where the model learns from labeled data, clustering works with unlabeled data. The goal is not to predict outcomes but to understand the data structure by identifying groups or clusters based on data similarity. The similarity is often calculated using distance measures such as Euclidean, Manhattan, or more sophisticated metrics tailored to the data type.

Let us understand the working mechanism of clustering by referring to the following figure:

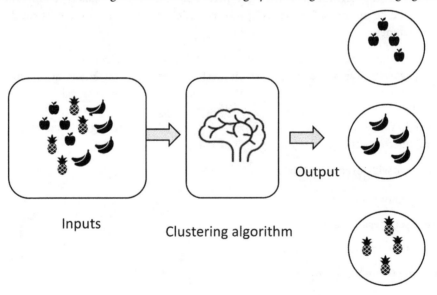

Figure 6.1 – Working mechanism of clustering

After understanding the working mechanism of clustering, let us understand the various scenarios where we can apply clustering.

When to use the clustering technique

Clustering can be used in the following scenarios:

- **Exploratory data analysis**: When you're in the initial stages of data analysis and want to understand the underlying structure or patterns in your data, clustering can reveal natural groupings, trends, and anomalies without the need for predefined labels.

- **No labeled data available**: In cases where you don't have labeled data (that is, you don't know the category/class of each data point beforehand), clustering is a go-to technique as it doesn't require such labels to group data points based on their similarity.

- **Discovering hidden patterns**: If the objective is to uncover hidden patterns or relationships within the data that are not known a priori, clustering can help identify these structures organically, providing insights that might not be evident through supervised methods such as classification.

- **Data preprocessing**: Clustering can be used as a preprocessing step to segment the data into more manageable groups. This segmentation can then inform or improve the performance of downstream tasks, including classification, by training separate classifiers on each cluster, for instance.

- **Feature engineering**: Clustering can also be employed in feature engineering to create new features that can enhance the performance of classification algorithms. For example, assigning cluster IDs as new features can provide additional context for supervised learning models.

- **Reduction of complexity in supervised tasks**: If a dataset is too complex or vast, clustering can be used to simplify it by grouping similar data points. This can make a subsequent classification task more manageable and improve the performance and interpretability of the classification models.

- **User or item profiling**: In systems such as recommendation engines, clustering can help profile users or items based on behaviors or attributes. These profiles can then enhance personalization strategies without needing explicit labels for each user or item.

As we can see, clustering is most beneficial when you need to explore data, find natural groupings, or work with unlabeled data. It's a powerful tool for discovering insights and patterns that are not immediately apparent, making it a valuable first step in many data analysis workflows, especially when complemented by classification and other machine learning techniques.

Some use cases of clustering in machine learning

Clustering, a key technique in unsupervised machine learning, is widely employed across various domains to discover inherent structures, group similar entities, and derive insights from unlabeled data. Here are some notable use cases where clustering is particularly beneficial:

- **Customer segmentation**: In marketing and sales, clustering helps identify distinct groups within a customer base based on purchasing behavior, preferences, demographics, and other factors. This enables businesses to tailor marketing strategies, product recommendations, and services to meet the specific needs of different customer segments.

- **Anomaly detection**: By grouping similar data points, clustering can highlight anomalies or outliers that do not fit into any cluster. This is crucial in fraud detection in banking and finance, in network security for identifying suspicious activities, and in manufacturing for detecting defective products.

- **Image segmentation**: In computer vision, clustering algorithms segment images into regions based on pixel similarities. This is essential for object recognition, medical imaging analysis (for example, identifying tumors in MRI scans), and enhancing image compression techniques.

- **Genomic data analysis**: In bioinformatics, clustering is used to group genes with similar expression patterns, which can indicate co-regulation or functional similarities. This is pivotal for understanding gene functions, identifying biomarkers for diseases, and exploring evolutionary relationships.

- **Recommendation systems**: Clustering can improve recommendation systems by grouping similar items or users. For instance, in e-commerce, products bought by similar customer groups can be recommended to others in the same cluster, enhancing the personalization of recommendations.

- **Text mining and Natural Language Processing (NLP)**: Clustering is applied to group similar documents, organize large sets of textual data, and extract thematic patterns in text corpora. Applications include news aggregation, sentiment analysis, and topic modeling.

- **Social network analysis**: Clustering algorithms can detect communities within social networks based on user interactions and shared interests. This helps in understanding social dynamics, information flow, and influence patterns within the network.

- **Market research**: Clustering provides insights into market structure and competitive landscapes by grouping similar products, services, or companies. This aids in market segmentation, trend analysis, and strategic planning.

- **Operational efficiency**: In logistics and supply chain management, clustering can optimize routes and schedules by grouping deliveries or pickups that are geographically close, leading to reduced costs and improved service levels.

- **Environmental studies**: Clustering aids in analyzing environmental data, such as clustering regions with similar climate patterns or pollution levels, contributing to ecological studies, conservation efforts, and climate change research.

In each of these use cases, clustering serves as a powerful tool to unveil patterns and similarities in data, leading to actionable insights, enhanced decision-making, and innovative solutions across a wide array of applications.

Pitfalls of clustering techniques

Clustering techniques, while powerful for exploratory data analysis and pattern recognition in unsupervised learning, come with several pitfalls that can impact their effectiveness and the interpretability of their results. Understanding these pitfalls is crucial for applying clustering methods appropriately and ensuring meaningful outcomes.

- **Choosing the number of clusters**: Many clustering algorithms, such as K-means, require the user to specify the number of clusters in advance. Determining the optimal number of clusters is not always straightforward and can significantly affect the results. Methods such as the elbow method, silhouette analysis, and gap statistics can help, but they might not always provide a clear answer.

- **Assumption of cluster shapes**: Different clustering algorithms make different assumptions about the shape and size of clusters. For example, K-means assumes clusters are spherical and of similar size, which might not hold true for all datasets. This can lead to suboptimal clustering when the actual clusters have irregular shapes or vastly different sizes.

- **Sensitivity to initial conditions**: Algorithms such as K-means are sensitive to the initial choice of cluster centers. Different initializations can lead to different clustering outcomes, potentially trapping the algorithm in local optima. Techniques such as K-means++ or multiple initializations with consensus clustering can mitigate this issue.

- **Noise and outliers**: Clustering algorithms can be sensitive to noise and outliers in the data. Noise can distort the true structure of the data, leading to clusters that don't accurately reflect the underlying patterns. Robust clustering techniques and preprocessing steps to remove outliers can help address this challenge.

- **Feature scaling**: The performance of many clustering algorithms is influenced by the scale of the data features. Variables on larger scales can unduly influence the distance metrics used for clustering, leading to biased results. Normalizing or standardizing data can help ensure that each feature contributes equally to the clustering process.

- **Interpretability**: The clusters produced by clustering algorithms can sometimes be difficult to interpret, especially in high-dimensional spaces. Understanding what each cluster represents in the context of the original data can be challenging, necessitating dimensionality reduction techniques or domain expertise for meaningful interpretation.

- **Validation of clusters**: Unlike supervised learning, where model performance can be evaluated against labeled test data, clustering lacks a clear objective measure of success. Validation is often subjective, relying on domain knowledge or indirect measures such as silhouette scores or cohesion and separation metrics.

- **Ignoring correlations among features**: Many clustering algorithms treat features as independent, ignoring potential correlations. This can lead to clusters that don't reflect complex relationships in the data. Dimensionality reduction techniques such as **Principal Component Analysis (PCA)** before clustering can help uncover these relationships.

- **High-dimensional data**: Clustering in high-dimensional spaces (the "curse of dimensionality") can be problematic. Distance metrics become less meaningful as dimensionality increases, and the sparsity of data points can make it difficult to identify dense regions or meaningful clusters. Techniques such as dimensionality reduction or using algorithms designed for high-dimensional data can help.

- **Algorithm complexity and scalability**: Some clustering algorithms can be computationally intensive, especially for large datasets or high-dimensional data. Scalability becomes a concern in big data applications, necessitating efficient algorithms or approximations.

By being aware of these pitfalls and carefully considering the choice of algorithm, parameters, and preprocessing steps, practitioners can mitigate some of the challenges associated with clustering techniques, leading to more reliable and interpretable clustering outcomes.

In the next section, you will understand more about clustering algorithms.

Learning clustering algorithms

There are different clustering algorithms available in Apache Spark. Let us now understand them in detail.

K-means

K-means is a widely used clustering algorithm in unsupervised machine learning that partitions a dataset into K distinct, non-overlapping clusters. It aims to minimize the variance within each cluster, essentially grouping data points based on their feature space similarities. The "means" in K-means refers to the centroids (mean points) of the K clusters, around which the data points are clustered.

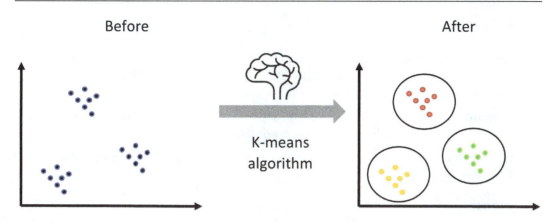

Figure 6.2 – Overview of K-means clustering

How K-means works

K-mean clustering involves the following steps:

1. **Initialization**: The algorithm starts by randomly selecting K points from the dataset as the initial centroids.

2. **Assignment step**: Each data point is assigned to the closest centroid. This forms K clusters.

3. **Update step**: The centroids are recalculated as the mean of all points assigned to each cluster.

4. **Iteration**: *Steps 2* and *3* are repeated until the centroids no longer significantly change, indicating convergence.

Applications

K-means can be applied in the following use cases:

- **Customer segmentation**: Businesses use K-means for market segmentation by clustering customers based on purchase history, behavior, and preferences to tailor marketing strategies and improve customer service.

- **Image compression**: K-means is used in image processing to reduce the number of colors in an image. Each cluster centroid represents a color, and pixels are assigned to the closest centroid color, significantly reducing the image size without substantial loss of quality.

- **Document clustering**: In text mining, K-means helps organize similar documents into groups, enhancing information retrieval systems by making it easier to find related documents.

Benefits

- **Simplicity and efficiency**: K-means is straightforward to understand and implement, making it accessible for a wide range of applications. It's computationally efficient for moderate to large datasets, making it suitable for quick exploratory data analysis.

- **Versatility**: The algorithm can be used with any type of feature if a distance measure can be defined, making it applicable to a wide variety of data types and domains.

- **Adaptability**: K-means can produce tighter clusters than hierarchical clustering, especially if the clusters are globular.

Limitations

- **Choosing K**: As discussed before, one of the biggest challenges is determining the optimal number of clusters, K. Inappropriate choices of K can lead to poor clustering performance. Methods such as the elbow method, silhouette analysis, and gap statistics can help but do not always provide a clear-cut answer.

- **Sensitivity to initial centroids**: The final clusters can depend heavily on the initial randomly chosen centroids. Poor initialization can lead to suboptimal solutions.

- **Assumption of spherical clusters**: K-means assumes that clusters are spherical and of a similar size, which might not always hold true. This can lead to poor performance if the actual clusters have different sizes, densities, or non-spherical shapes.

K-means remains a popular clustering algorithm due to its simplicity, efficiency, and effectiveness in many practical applications. However, understanding its limitations and considering the nature of the dataset is crucial in ensuring meaningful clustering outcomes.

Code example

Having explored the clustering concepts, we will walk through the code example of K-means clustering.

The following code imports the required libraries and creates a Spark session:

```
from sklearn.datasets import make_blobs
import numpy as np
from pyspark.sql.functions import monotonically_increasing_id
from pyspark import SparkFiles
from pyspark.sql import SparkSession
from pyspark.ml.feature import (
    VectorAssembler, StandardScaler)
from pyspark.ml.clustering import KMeans
from pyspark.ml.evaluation import ClusteringEvaluator
import matplotlib.pyplot as plt
import pyspark
```

```
spark = pyspark.sql.SparkSession.builder.appName(
    "K-means Clustering"
).getOrCreate()
```

The following code creates synthetic data:

```
features, true_labels = make_blobs(
    ...:         n_samples=200,
    ...:         centers=3,
    ...:         cluster_std=2.75,
    ...:         random_state=42
    ...: )
```

The following code will output the generated data through a Spark DataFrame:

```
df1 = spark.createDataFrame(features,['feature1','feature2'])
df1 = df1.withColumn("id1", monotonically_increasing_id())
df2 = spark.createDataFrame(true_labels,['label'])
df2 = df2.withColumn("id1", monotonically_increasing_id())
df = df1.join(df2,on="id1")
df.show(5)
```

We will get the following results:

```
+---+------------------+-------------------+-----+
|id1|          feature1|           feature2|label|
+---+------------------+-------------------+-----+
|  0|  9.770758741876183| 3.2762102164481477|    1|
|  1| -9.713496659299548| 11.274518015230187|    0|
|  2| -6.913305816480285| -9.347559114861983|    2|
|  3|-10.861859130268911|-10.750634972811474|    2|
|  4|  -8.50038027274785| -4.543703826468128|    2|
+---+------------------+-------------------+-----+
only showing top 5 rows
```

Figure 6.3 – Output of the generated data

The next code snippet assembles features into a vector column and then scales those features using standardization:

```
assembler = VectorAssembler(
    inputCols=["feature1", "feature2"],
    outputCol="features")
data_df = assembler.transform(df)
scaler = StandardScaler(inputCol="features",
    outputCol="scaled_features")
```

```
scaler_model = scaler.fit(data_df)
data_df = scaler_model.transform(data_df)
data_df.show(5)
```

The resulting DataFrame (data_df) contains both the original features and the scaled features:

```
+---+-------------------+-------------------+-----+-------------------+-------------------+
|id1|           feature1|           feature2|label|           features|    scaled_features|
+---+-------------------+-------------------+-----+-------------------+-------------------+
|  0|  9.770758741876183| 3.2762102164481477|    1|[9.77075874187618...|[1.82968666370121...|
|  1| -9.713496659299548| 11.274518015230187|    0|[-9.7134966592995...|[-1.8189636818331...|
|  2| -6.913305816480285| -9.347559114861983|    2|[-6.9133058164802...|[-1.2945958229721...|
|  3|-10.861859130268911|-10.750634972811474|    2|[-10.861859130268...|[-2.0340077284354...|
|  4|  -8.50038027274785| -4.543703826468128|    2|[-8.5003802727478...|[-1.5917937216868...|
+---+-------------------+-------------------+-----+-------------------+-------------------+
only showing top 5 rows
```

Figure 6.4 – DataFrame showing original and scaled features

The following code snippet applies the K-means algorithm to show the cluster for each data point:

```
kmeans = KMeans(k=3, featuresCol="scaled_features",
    predictionCol="cluster")
model = kmeans.fit(data_df)
predictions = model.transform(data_df)
predictions.show(5)
```

It will produce the following output:

```
+---+-------------------+-------------------+-----+-------------------+-------------------+-------+
|id1|           feature1|           feature2|label|           features|    scaled_features|cluster|
+---+-------------------+-------------------+-----+-------------------+-------------------+-------+
|  0|  9.770758741876183| 3.2762102164481477|    1|[9.77075874187618...|[1.82968666370121...|      1|
|  1| -9.713496659299548| 11.274518015230187|    0|[-9.7134966592995...|[-1.8189636818331...|      0|
|  2| -6.913305816480285| -9.347559114861983|    2|[-6.9133058164802...|[-1.2945958229721...|      2|
|  3|-10.861859130268911|-10.750634972811474|    2|[-10.861859130268...|[-2.0340077284354...|      2|
|  4|  -8.50038027274785| -4.543703826468128|    2|[-8.5003802727478...|[-1.5917937216868...|      2|
+---+-------------------+-------------------+-----+-------------------+-------------------+-------+
only showing top 5 rows
```

Figure 6.5 – DataFrame output showing cluster assignment

The next code snippet evaluates the clustering output:

```
evaluator = ClusteringEvaluator(predictionCol='cluster',
    featuresCol='scaled_features', \
    metricName='silhouette', \
    distanceMeasure='squaredEuclidean')
silhouette_score = evaluator.evaluate(predictions)
print(f"Silhouette Score: {silhouette_score:.4f}")
```

We receive the Silhouette Score: 0.7851 output.

Now, let's find the optimal number of clusters:

```
wssse_values =[]
evaluator = ClusteringEvaluator(predictionCol='cluster',
    featuresCol='scaled_features', \
    metricName='silhouette', \
    distanceMeasure='squaredEuclidean')
for i in range(2,8):
    KMeans_mod = KMeans(k=i, featuresCol="scaled_features",
        predictionCol="cluster")
    KMeans_fit = KMeans_mod.fit(data_df)
    output = KMeans_fit.transform(data_df)
    score = evaluator.evaluate(output)
    wssse_values.append(score)
    print("Silhouette Score:",score)
```

This will generate the output as shown:

```
Silhouette Score: 0.7178670008193637
Silhouette Score: 0.7851032706829328
Silhouette Score: 0.6726486143551612
Silhouette Score: 0.5810267089116743
Silhouette Score: 0.5446619451331941
Silhouette Score: 0.5228633085259329
```

Figure 6.6 – Silhouette scores output

The following code snippet plots the graph showing the WSSSE values for a range of cluster sizes:

```
# Plotting WSSSE values
plt.plot(range(1, 7), wssse_values)
plt.xlabel('Number of Clusters (K)')
plt.ylabel('Within Set Sum of Squared Errors (WSSSE)')
plt.title('Elbow Method for Optimal K')
plt.grid()
plt.show()
```

The graph for finding the optimal number of clusters using the elbow method is as shown:

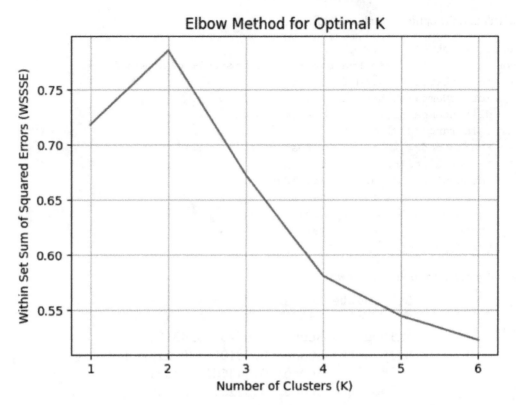

Figure 6.7 – Graph for finding the optimal K

Next, we will plot all the data points along with cluster centers to visualize the clusters:

```
plt.figure(figsize=(8, 6))
for i in range(3):
    cluster_data = predictions.filter(
        predictions["cluster"] == i
    ).select("scaled_features").collect()
    cluster_points = [point[0] for point in cluster_data]
    plt.scatter(*zip(*cluster_points), label=f"Cluster {i}")
plt.scatter(*zip(*model.clusterCenters()), c='black',
    s=200, alpha=0.5, label="Cluster Centers")
plt.title("K-Means Clustering")
plt.xlabel("PCA Component 1")
plt.ylabel("PCA Component 2")
plt.legend()
plt.show()
```

Here's the output generated:

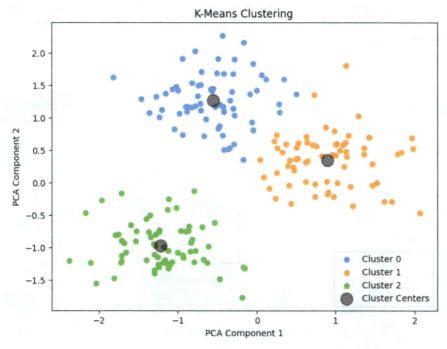

Figure 6.8 – Cluster visualization

Latent Dirichlet allocation (LDA)

LDA is a generative statistical model that explains sets of observations using unobserved groups. It helps us understand why certain parts of the data exhibit similarities. In the context of text analysis, LDA assumes that documents are produced from a mixture of topics, where a topic is understood as a distribution over a fixed vocabulary. The key idea is that documents exhibit multiple topics to different extents, and LDA aims to uncover these latent topic structures.

The LDA model

LDA clusters documents as random mixtures over latent topics, where each topic is represented by a distribution over words. LDA assumes the following generative process for each document in a corpus.

Choose a distribution over topics

For each word in the document, do the following:

- Choose a topic from the document's distribution over topics
- Choose a word from the topic's distribution over words

The distributions are determined by Dirichlet distributions, parameterized by hyperparameters usually fixed before inference begins. The inference process aims to backtrack from the documents to find a set of topics that are likely to have generated the corpus. The following picture explains the concept of **Latent Dirichlet Allocation (LDA)**:

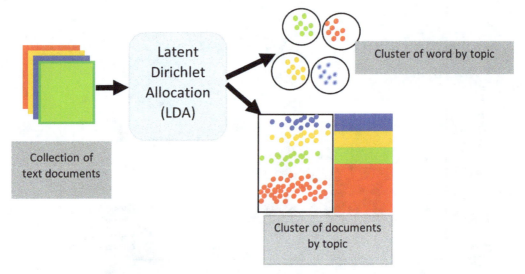

Figure 6.9 – Overview of LDA

Applications of LDA

LDA can be used in the following use cases:

- **Topic modeling in text data**: LDA is extensively used for discovering the underlying thematic structure in large collections of texts, making it easier to manage and navigate through large datasets by summarizing them into topics

- **Content recommendation**: By understanding the topics present in articles or products, systems can recommend new content to users based on their past preferences for certain topics

- **Information retrieval**: This involves enhancing search engines to return documents that are not just keyword matches but also thematically relevant, by indexing documents based on their topic distributions

Benefits

- **Unsupervised learning**: As an unsupervised technique, LDA does not require labeled data, making it valuable for exploring large datasets where manual labeling is infeasible

- **Dimensionality reduction**: LDA reduces the dimensionality of text data by representing documents as mixtures of a smaller number of topics, simplifying subsequent analysis and visualization

- **Interpretability**: The topics produced by LDA are often coherent and interpretable, allowing humans to understand and label the themes within the data, providing insights into the structure and content of large text corpora

Limitations

- **Choosing the number of topics**: Like other clustering techniques, determining the optimal number of topics (K) is challenging. Various heuristic methods exist, but there's no one-size-fits-all solution, and the choice can significantly affect the results.

- **Assumption of topic independence**: LDA assumes that topics are independent and do not influence each other, which might not always hold true in real-world data where topics can be hierarchical or correlated.

- **Handling of polysemy and synonymy**: LDA treats each word as a distinct token and does not account for polysemy (words with multiple meanings) or synonymy (different words with similar meanings), which can lead to less meaningful topics.

Overall, LDA is a powerful tool for uncovering latent thematic structures in large text corpora, offering valuable insights for exploratory data analysis, content recommendation, and information retrieval. Its unsupervised nature and the ability to produce interpretable topics make it widely applicable, although care must be taken in model specification, hyperparameter selection, and result interpretation to address its inherent limitations.

Code example

First, we will import the required libraries and create the Spark session:

```
from sklearn.datasets import fetch_20newsgroups
from pyspark.sql import SparkSession
from pyspark.sql.functions import udf,expr
from pyspark.sql.types import StringType
from pyspark.ml.feature import (
    StopWordsRemover, Tokenizer,CountVectorizer,IDF)
import re
# Create a Spark session
spark = SparkSession.builder.appName(
    "NewsgroupsPreprocessing"
).getOrCreate()
spark.sparkContext.setLogLevel("ERROR")
num_features = 8000  #vocabulary size
num_topics = 20
```

Then we will load the `20newsgroups` dataset:

```
# Load the 20-Newsgroups dataset
newsgroups = fetch_20newsgroups(
    subset="all", remove=("headers", "footers", "quotes"))

df = spark.createDataFrame(
    [(doc,) for doc in newsgroups.data], ["doc"])
```

The following code snippet uses regex functions to remove non-alphabetic characters:

```
from pyspark.sql.functions import regexp_replace
# Define a UDF to apply regex and remove non-alphabetic characters
def clean_text(doc):
    cleaned_doc = re.sub(r"[^A-Za-z]", " ", doc)
    return " ".join(cleaned_doc.split())  # Remove extra spaces

# Register the UDF
clean_text_udf = udf(clean_text, StringType())

# Apply regex
df_cleaned = df.withColumn("cleaned_doc", clean_text_udf("doc"))
# Replace newline characters with a space
df_cleaned = df.withColumn("cleaned_text",
    regexp_replace("doc", "\n", " "))
df_cleaned.select("cleaned_text").show(1,truncate=False)
```

Then we use the `tokenizer`, `stopwordsremover`, and `countvectorizer` functions to generate the features:

```
# Apply Tokenizer
tokenizer = Tokenizer(inputCol="cleaned_text", outputCol="tokens")
df_tokenized = tokenizer.transform(df_cleaned)
# Apply StopWordsRemover
stopwords_remover = StopWordsRemover(
    inputCol="tokens", outputCol="filtered_doc")
df_filtered = stopwords_remover.transform(
    df_tokenized
).select("filtered_doc")
# Filter array elements with at least 4 characters
df_filtered = df_filtered.withColumn(
    "filtered_array",
    expr("filter(filtered_doc, x -> len(x) >= 4)"))
df_filtered = df_filtered.select("filtered_array")
df_filtered.show(1,truncate=False)
```

```
# Apply CountVectorizer
count_vec = CountVectorizer(inputCol="filtered_array",
    outputCol="count_vec" ,vocabSize=num_features, minDF=2.0)
count_vec_model = count_vec.fit(df_filtered)
```

We will get the following output:

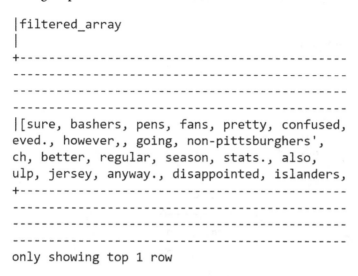

```
|filtered_array
|
+---------------------------------------------
----------------------------------------------
----------------------------------------------
----------------------------------------------
|[sure, bashers, pens, fans, pretty, confused,
eved., however,, going, non-pittsburghers',
ch, better, regular, season, stats., also,
ulp, jersey, anyway., disappointed, islanders,
+---------------------------------------------
----------------------------------------------
----------------------------------------------
----------------------------------------------
only showing top 1 row
```

Figure 6.10 – Spark DataFrame output

The next code snippet transforms the data to vectorized features using the IDF function:

```
vocab = count_vec_model.vocabulary
newsgroups = count_vec_model.transform(df_filtered)
#newsgroups = newsgroups.drop('filtered_array')
 # Apply IDF
idf = IDF(inputCol="count_vec", outputCol="features")
newsgroups = idf.fit(newsgroups).transform(newsgroups)
newsgroups = newsgroups.drop('tf_features')
```

Now it's time to generate the LDA model:

```
from pyspark.ml.clustering import LDA
lda = LDA(k=num_topics, featuresCol="features", seed=0)
model = lda.fit(newsgroups)
transformed_data = model.transform(newsgroups)
transformed_data.show(5)
topics = model.describeTopics()
topics.show(5)
```

```
model.topicsMatrix()

topics_rdd = topics.rdd
topics_words = topics_rdd\
    .map(lambda row: row['termIndices'])\
    .map(lambda idx_list: [vocab[idx] for idx in idx_list])\
    .collect()
```

The following code snippet shows the topics:

```
for idx, topic in enumerate(topics_words):
    print ("topic: ", idx)
    print ("----------")
    for word in topic:
        print (word)  # word
    print ("----------")
```

```
----------
topic:  1
----------
armenian
armenians
jews
turkish
people
rights
jewish
israeli
right
government
----------
```

Figure 6.11 – Keywords contained in topic: 1

Bisecting K-means

Bisecting K-means is a clustering algorithm that is a variant of the traditional K-means algorithm, designed to produce a hierarchical decomposition of the dataset. It combines aspects of K-means with a divisive (top-down) hierarchical clustering approach, making it distinct in its operation and application.

How bisecting K-means works

The process of bisecting K-means involves repeatedly splitting clusters until the desired number of clusters is achieved. The steps can be outlined as follows:

1. **Initialization**: Start with all data points in a single cluster.

2. **Bisection**: Select a cluster to split. This can be the largest cluster, the one with the highest **Sum of Squared Errors (SSE)**, or one that is chosen based on other criteria.

3. **K-means**: Apply the K-means algorithm with **K=2** on the selected cluster to split it into two sub-clusters.

4. **Iteration**: Repeat the bisection step, choosing from the newly created clusters and the existing ones, until the desired number of clusters is reached.

Throughout this process, the algorithm iteratively refines the clusters to improve their coherence, using the standard K-means optimization criteria of minimizing the within-cluster variance. The following diagram shows this process:

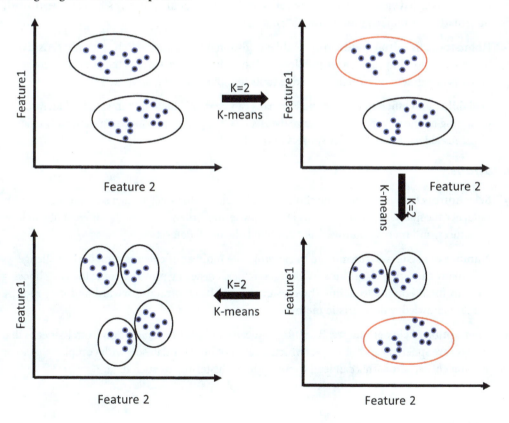

Figure 6.12 – Working of bisecting K-means

Applications

This technique can be applied in the following contexts:

- **Document clustering**: In text mining and information retrieval, bisecting K-means is used to organize large collections of documents into hierarchically structured groups based on their content similarity, enhancing document navigation and topic discovery

- **Image segmentation**: The algorithm can segment digital images into coherent regions, aiding in object recognition, compression, and detailed image analysis

- **Market research**: Bisecting K-means helps in customer segmentation, dividing customers into distinct groups based on purchasing patterns, preferences, or demographics, thereby enabling targeted marketing strategies

Benefits

- **Hierarchical structure**: The divisive approach provides a hierarchical clustering structure, offering a detailed view of data grouping at various levels of granularity, which can be particularly informative for understanding complex datasets

- **Robustness to initial conditions**: Unlike traditional K-means, which is highly sensitive to initial centroids and can converge to local minima, bisecting K-means tends to be more stable and less dependent on initialization due to its iterative bisection approach

- **Scalability and efficiency**: Bisecting K-means can be more efficient than standard hierarchical clustering methods, especially for large datasets, as it applies the more computationally efficient K-means algorithm on smaller subsets of the data

Limitations

- **Selection of clusters to split**: The choice of which cluster to bisect at each step can significantly influence the final clustering outcome. Different selection criteria can lead to different hierarchical structures, and there's no universally optimal rule for all datasets.

- **Number of clusters**: Determining the appropriate number of clusters remains a challenge, as with many clustering algorithms. While hierarchical clustering provides some insight, choosing a cut in the hierarchy to define the clusters is not always straightforward and may require domain knowledge or heuristic methods.

- **Assumption of cluster Shapes**: Bisecting K-means inherits the assumption from K-means that clusters are spherical and isotropic, which might not hold for datasets with complex shapes or varying cluster sizes. This can lead to suboptimal clustering for such data.

As we can see, bisecting K-means offers a unique approach to clustering by combining the efficiency of K-means with a hierarchical structure, providing a versatile tool for data analysis. Its ability to generate hierarchical clustering, along with its robustness and efficiency, makes it suitable for a wide range of applications. However, careful consideration of its limitations and implementation details is crucial to leverage its full potential and achieve meaningful clustering outcomes.

Gaussian Mixture Model (GMM)

The **Gaussian Mixture Model** (**GMM**) is a probabilistic model with the assumption that all the data points are generated from a mixture of a finite number of Gaussian distributions with unknown parameters. GMMs are a type of mixture model that provide a method for representing an underlying, hidden data structure; they are particularly well-suited to identifying subpopulations within a given dataset, making them powerful tools for clustering, density estimation, and pattern recognition.

Understanding Gaussian Mixture Models

A GMM is characterized by two main components: the mixing coefficients and the components' Gaussian distributions. Each component distribution is defined by its mean and covariance, which determines the location, shape, and orientation of the distribution in the data space. The mixing coefficients represent the weights of each Gaussian component in the mixture, indicating the proportion of the population that belongs to each component.

GMMs belong to the family of **Expectation-Maximization** (**EM**) algorithms, where the **Expectation** step estimates the membership weights of each data point in each cluster (based on current parameter estimates), and the **Maximization** step updates the parameters of the Gaussians to maximize the likelihood of the data given these memberships.

In the following figure, three clusters are represented by three Gaussian curves. The centroid of each cluster is shown using the vertical dotted line at the center of each curve.

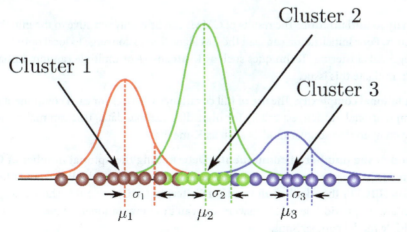

Figure 6.13 – Overview of GMM

Applications

- **Image segmentation**: GMMs are used in computer vision to segment images into regions with similar textures or colors. Each Gaussian component can model the pixel intensities of different image regions, allowing for detailed and flexible image segmentation.

- **Speaker verification**: In the field of speech processing, GMMs can model the voice characteristics of individual speakers. By learning the distribution of a speaker's voice features, GMMs can verify a speaker's identity by comparing the likelihood of observed features under their model.

- **Financial market analysis**: GMMs can model the behavior of financial assets, capturing the multimodal distributions of asset returns. This helps in identifying different market regimes, such as high volatility or stable growth periods, and can inform risk management and investment strategies.

Benefits

- **Flexibility in shape of clusters**: Unlike K-means, which assumes spherical clusters, GMMs can accommodate ellipsoidal clusters of different sizes and orientations due to their parametric nature, providing a more nuanced clustering solution.

- **Soft clustering**: GMMs offer probabilistic cluster assignments, meaning data points can belong to multiple clusters with varying degrees of membership. This soft clustering approach captures the uncertainty in data grouping, which can be valuable in many real-world scenarios where data inherently belongs to overlapping categories.

- **Density estimation**: Beyond clustering, GMMs are effective in estimating the density of complex datasets. They can model the distribution of data points in high-dimensional spaces, making them useful for anomaly detection, generative models, and other applications where understanding the underlying data distribution is crucial.

Limitations

- **Sensitivity to initialization**: The results of GMMs can be highly sensitive to the initial choice of parameters. Poor initialization can lead the EM algorithm to converge to local optima, resulting in suboptimal clustering. Techniques such as **K-Means**++ or multiple random initializations can help mitigate this issue.

- **Computational complexity**: The use of full covariance matrices for each component increases the computational burden, especially for high-dimensional data. This can make GMMs less scalable compared to simpler models such as K-means.

- **Selection of the number of components**: Determining the optimal number of Gaussian components in the mixture is a non-trivial task. Criteria such as the **Bayesian Information Criterion** (**BIC**) or the **Akaike Information Criterion** (**AIC**) can guide this choice, but they may not always provide clear-cut answers and can be computationally expensive to compute for multiple model comparisons.

GMMs offer a powerful and flexible framework for clustering and density estimation, capable of capturing complex data structures. Their probabilistic foundation and the ability to model different cluster shapes make them a valuable tool in a wide array of applications. However, careful attention to their limitations and implementation details is essential to leverage their full potential and achieve meaningful insights from data.

Code example for clustering using a Gaussian Mixture Model

We start with importing the required libraries and creating synthetic data:

```
import numpy as np
from sklearn.datasets import make_blobs
import matplotlib.pyplot as plt
X, y = make_blobs(n_samples=300, centers=3, random_state=42)
plt.scatter(X[:, 0], X[:, 1], c=y, cmap='viridis')
plt.xlabel('Feature 1')
plt.ylabel('Feature 2')
plt.title('Synthetic Data with 3 Clusters')
plt.show()
```

Here's the output:

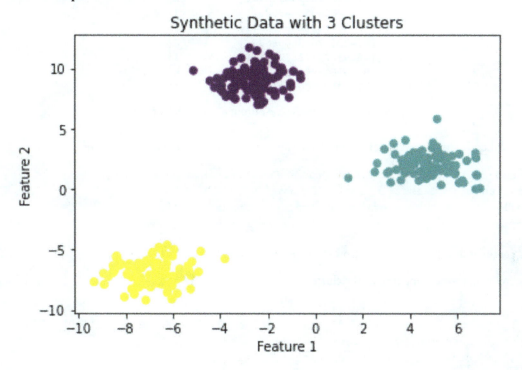

Figure 6.14 – Clustering visualization

The next code snippet imports all the required libraries, creates a Spark session, fits the data to the model, and shows the prediction output:

```
from pyspark.ml.clustering import GaussianMixture
from pyspark.ml.linalg import Vectors
from pyspark.sql import SparkSession
spark = SparkSession.builder.appName(
    "GMMClustering"
).getOrCreate()
data = [(Vectors.dense(X[i]),) for i in range(len(X))]
df = spark.createDataFrame(data, ["features"])
gm = GaussianMixture(k=3, tol=0.0001, seed=10)
model = gm.fit(df)
transformed = model.transform(df).select("features", "prediction")
transformed.show()
```

The output is as shown:

```
+--------------------+----------+
|            features|prediction|
+--------------------+----------+
|[-7.3389880906915...|         1|
|[-7.7400405564352...|         1|
|[-1.6866527109495...|         2|
|[4.42219763300088...|         0|
|[-8.9177517263291...|         1|
|[5.49753845943012...|         0|
|[-2.3360166972015...|         2|
```

Figure 6.15 – GMM prediction output along with features

The next code snippet evaluates the model performance and prints the metrics output:

```
from pyspark.ml.evaluation import ClusteringEvaluator
evaluator = ClusteringEvaluator()
silhouette_score = evaluator.evaluate(transformed)
print(f"Silhouette Score: {silhouette_score}")
```

We get the following output: Silhouette Score: 0.9696628776463786.

The following code snippet plots the cluster:

```
transformed_pd = transformed.toPandas()
plt.scatter(transformed_pd["features"].apply(lambda x: x[0]),
    transformed_pd["features"].apply(lambda x: x[1]),
    c=transformed_pd["prediction"], cmap='viridis')
plt.xlabel('Feature 1')
```

```
plt.ylabel('Feature 2')
plt.title('Clusters from GMM')
plt.show()
```

Let's see the output:

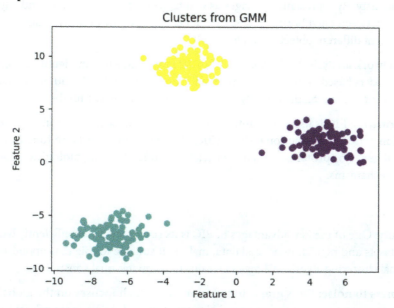

Figure 6.16 – Visualization of clusters using GMM

Power Iteration Clustering (PIC)

Power Iteration Clustering (PIC) is a scalable and efficient graph clustering algorithm that employs the concept of power iteration, a method commonly used in eigenvalue solvers, to project high-dimensional data points into a low-dimensional space. This projection is based on the similarity matrix of the data, representing the pairwise similarities between data points as edges in a graph, thereby transforming the clustering problem into a graph partitioning problem.

How PIC works

PIC starts by constructing a similarity matrix that represents the dataset as a graph, where nodes correspond to data points and edges reflect the similarities between these points. The algorithm then performs power iterations on this graph to compute the principal eigenvector of the modified normalized Laplacian matrix of the graph. The resulting eigenvector reflects the steady-state distribution of a random walk on the graph, capturing the essential structure of the data. Data points are then mapped to a low-dimensional space based on their components in the principal eigenvector, and traditional clustering techniques, such as K-means, can be applied in this reduced space to identify clusters.

Applications

The PIC technique is used in the following contexts:

- **Image segmentation**: PIC can be applied to segment images into coherent regions based on pixel similarity. By representing images as graphs where pixels are nodes and edges encode similarity between neighboring pixels, PIC can effectively partition the image into segments that represent different objects or regions.

- **Social network analysis**: In the analysis of social networks, PIC can identify communities by clustering users based on their interactions and connections. The algorithm can handle large-scale networks, revealing the underlying community structure within the network.

- **Bioinformatics**: PIC is used in bioinformatics for clustering genes or proteins based on their expression patterns or functional similarities. This helps in identifying functionally related groups of genes or proteins, which is crucial for understanding biological processes and disease mechanisms.

Benefits

- **Scalability**: One of the key advantages of PIC is its scalability. It can efficiently handle large-scale datasets and high-dimensional data, making it suitable for big data applications where traditional clustering algorithms may falter due to computational constraints.

- **Robustness to noise**: PIC's graph-based approach, which focuses on the global structure of the data represented by the principal eigenvector, tends to be more robust to noise and outliers compared to methods that rely on local information or distance metrics in the original high-dimensional space.

- **Flexibility in defining similarity**: The algorithm allows for flexibility in defining the similarity measure between data points, enabling it to be tailored to specific types of data or applications. This can include linear similarities for Euclidean spaces or more complex measures for structured data such as graphs or manifolds.

Limitations

- **Dependence on similarity measure**: The effectiveness of PIC heavily depends on the choice of the similarity measure and the construction of the similarity matrix. An inappropriate similarity measure can lead to poor clustering results, as it may not capture the true relationships between data points.

- **Selection of parameters**: PIC involves choosing parameters such as the number of clusters and the scaling parameter for the similarity matrix. Inappropriate choices can affect the quality of the clustering, and there may not be straightforward methods for selecting these parameters optimally.

- **Interpretation of results**: The low-dimensional representation and the resulting clusters obtained from PIC may not always be easy to interpret, especially in applications where the original feature space has a clear and meaningful structure. The abstraction of a graph representation and subsequent eigenvector-based projection may obscure the intuitive understanding of cluster membership.

PIC offers a powerful approach to clustering large-scale and high-dimensional data by leveraging graph representations and eigenvector computations. Its scalability and robustness to noise make it suitable for a wide range of applications, from image analysis to social network exploration. However, its effectiveness is contingent on the careful selection of similarity measures, parameters, and the interpretability of the clustering outcomes, necessitating a thoughtful and informed implementation strategy.

Code example using Power Iteration Clustering

Now that we have learned about PIC, let's walk through a code example:

```
from sklearn.datasets import load_iris
# Load the Iris dataset
iris = load_iris()
# Features (sepal length, sepal width, petal length, petal width)
X = iris.data
# Target labels (0: Setosa, 1: Versicolour, 2: Virginica)
y = iris.target
import seaborn as sns
import matplotlib.pyplot as plt
# Load the Iris dataset
iris = sns.load_dataset("iris")
# Scatter plot of Sepal Length vs. Petal Length
sns.scatterplot(x="sepal_length", y="petal_length",
    hue="species", data=iris)
plt.xlabel("Sepal Length (cm)")
plt.ylabel("Petal Length (cm)")
plt.title("Iris Dataset: Sepal Length vs. Petal Length")
plt.show()
```

This code will generate the following output:

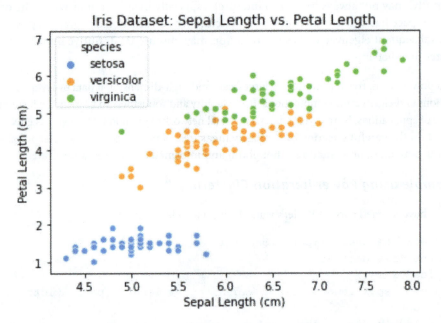

Figure 6.17 – Visualization of the dataset

The following code snippet imports all the required libraries, creates a Spark session, and builds the PIC clustering model:

```
from pyspark.ml.clustering import PowerIterationClustering
from pyspark.sql import SparkSession
# Create a Spark session
spark = SparkSession.builder.appName(
    "PICClustering"
).getOrCreate()
# Convert the scikit-learn data to a DataFrame
data = [(i, j, 1.0)
    for i in range(len(X))
    for j in range(len(X)) if i != j]
df = spark.createDataFrame(data, ["src", "dst", "weight"])
# Create a PowerIterationClustering model
pic = PowerIterationClustering(k=3, weightCol="weight")
# Run the PIC algorithm and get cluster assignments
assignments = pic.assignClusters(df)
# Show the cluster assignments
assignments.show()
```

We will get the following output:

```
+---+--------+
| id|cluster|
+---+--------+
| 96|      0|
| 56|      2|
|112|      2|
|120|      2|
```

After learning several different clustering techniques and algorithms, let us look at how to evaluate the model performance.

Evaluating the model performance

In this section, we will look at several evaluation metrics to evaluate the performance of the clustering model.

Evaluation clustering algorithms

Evaluating clustering algorithms involves measuring how well the algorithm has identified distinct clusters that are meaningful and well-separated. The choice of evaluation metrics can depend on the availability of ground truth labels and the specific characteristics of the clustering technique. Here's how various evaluation metrics apply to the clustering techniques you've mentioned.

When evaluating clustering results, it's essential to consider the context of the application and the characteristics of the data. Often, a combination of metrics is used to provide a comprehensive assessment of the clustering quality. Furthermore, visual inspections, domain-specific validations, and stability assessments across multiple runs can complement these quantitative metrics for a more robust evaluation.

Selecting the evaluation metrics

When evaluating clustering results, it's essential to consider the context of the application and the characteristics of the data. Often, a combination of metrics is used to provide a comprehensive assessment of the clustering quality. Furthermore, visual inspections, domain-specific validations, and stability assessments across multiple runs can complement these quantitative metrics for a more robust evaluation.

Selecting the right evaluation metrics for clustering techniques discussed in this chapter involves understanding both the nature of the clustering algorithm and the specific characteristics of the data. Different metrics can provide various insights into the quality of the clustering, such as cohesion, separation, relevance to predefined categories, or the match with known ground truth if available.

Common evaluation metrics

Let us look at different evaluation metrics:

- **Silhouette score:** As we saw earlier, this measures the similarity of an object to its own cluster compared to other clusters. The silhouette score ranges from -1 to 1. A high value indicates that the object is well matched to its own cluster and poorly matched to neighboring clusters. It's applicable to K-means, bisecting K-means, GMM, and PIC, especially when there's no ground truth available.

- **Davies-Bouldin Index:** A metric that evaluates intra-cluster similarity and inter-cluster differences. Lower values indicate better clustering. It's suitable for methods such as K-means, bisecting K-means, and GMM, where defining clusters as compact and well-separated is desired.

- **Calinski-Harabasz index:** Also known as the **Variance Ratio Criterion** (**VRC**), this metric evaluates the ratio of the sum of between-clusters dispersion and intra-cluster dispersion for all clusters. Higher values typically indicate better-defined clusters and can be used across K-means, bisecting K-means, GMM, and PIC.

- **Adjusted Rand Index** (**ARI**): This measures the similarity between two clusterings by accounting for all pairs of samples and counting pairs assigned in the same or different clusters in the predicted and true clustering. It's particularly useful when you have a ground truth. ARI can be used for all the mentioned clustering techniques when labeled data is available for comparison.

- **Normalized Mutual Information** (**NMI**): This is a normalization of the **Mutual Information** (**MI**) score to scale the results between 0 (no MI) and 1 (perfect correlation). Like ARI, it's used when ground truth labels are available, making it applicable for evaluating all mentioned clustering methods in a supervised context.

- **Perplexity:** Often used in topic models such as LDA, perplexity measures how well a probability distribution predicts a sample. Lower perplexity indicates better generalization performance

- **Log-likelihood:** This indicates how likely the observed data are given the model parameters. Higher values suggest a better model but need to be carefully interpreted, especially when comparing models with different numbers of parameters. This is used in GMM.

- **Dunn Index:** This measures the ratio between the smallest distance between observations not in the same cluster to the largest intra-cluster distance. Higher values indicate better clustering.

To use the silhouette score in Spark for evaluating your clustering model, you can follow these general steps:

1. Import `ClusteringEvaluator` from Spark MLlib.

2. Instantiate `ClusteringEvaluator`, specifying the metric name as `silhouette` if necessary (this is the default metric).

3. Evaluate your model by passing the predictions DataFrame to the evaluate method of `ClusteringEvaluator`. The predictions DataFrame should include a prediction column indicating the cluster assigned to each instance.

While Spark's built-in support for clustering evaluation metrics is currently focused on the silhouette score, the extensible nature of Spark allows for the implementation of additional metrics as needed to suit your specific evaluation criteria and clustering analysis goals.

Selection criteria and conditions

Let us understand the context for using different evaluation metrics:

- **Absence of ground truth**: In purely unsupervised scenarios where no external labels are available, intrinsic metrics such as the silhouette score, Davies-Bouldin Index, or Calinski-Harabasz index are more appropriate as they rely solely on the clustering results and the data itself.

- **Presence of ground truth**: When external labels or ground truth are available, even just for a subset of the data, extrinsic metrics such as ARI or NMI provide a direct measure of how well the clustering matches the known labels.

- **Nature of clusters**: Consider the expected cluster shapes and distributions. Metrics that assume spherical clusters (such as some interpretations of the silhouette score) may not be appropriate for evaluating clusters produced by GMM, which can capture more complex, elliptical distributions.

- **Scalability and computational cost**: For very large datasets, computationally intensive metrics may not be practical. Choose metrics that can be efficiently computed at scale, or consider using a sample of the data for evaluation.

- **Interpretability**: The choice of metric should also align with the stakeholders' needs for interpretability. For example, the silhouette score can provide more intuitive insights into cluster cohesion and separation for non-technical audiences.

- **Multiple metrics**: Often, relying on a single metric can be misleading. Consider using a combination of metrics to get a more comprehensive view of the clustering performance. Each metric can highlight different aspects of the clustering quality, providing a more nuanced understanding.

To summarize, selecting evaluation metrics for clustering techniques involves a balance between the methodological characteristics of the algorithms, the nature of the data, the computational context, and the specific objectives of the clustering task. Understanding each metric's strengths and limitations is crucial in making informed choices aligning with the analysis's goals.

Improving the model performance

Improving the performance of clustering models involves a multifaceted approach. These strategies span from data preprocessing to algorithm-specific adjustments and post-clustering evaluation.

General strategies for all models

The strategies for improving model performance can be broadly categorized as follows:

- **Data preprocessing**:

    - **Normalization/standardization**: Ensure all features contribute equally by scaling the data, especially for algorithms such as K-means and GMM, which are sensitive to the scale of the data

    - **Noise reduction**: Remove outliers and noise to prevent them from skewing the clustering

    - **Dimensionality reduction**: Use algorithms such as PCA or t-SNE to reduce the dimensionality, which can help in mitigating the curse of dimensionality and improving clustering quality

- **Feature selection and engineering**:

    - Identify and select the most relevant features for the clustering task

    - Construct new features that might better capture the inherent structures in the data

- **Hyperparameter tuning**:

    - Experiment with different hyperparameters, such as the number of clusters in K-means or the Dirichlet distribution parameters in LDA, to find the optimal settings

Model-specific strategies

We can focus on strategies based on clustering techniques as follows:

- **K-means**:

    - **Smart initialization**: Use methods such as K-means++ for smarter centroid initialization to avoid poor local minima

    - **Multiple runs**: Perform multiple initializations and select the run with the best convergence to mitigate the effects of random initialization

    - **Elbow method**: Use the elbow method to help determine the optimal number of clusters, balancing between model complexity and explained variance

- **LDA**:

    - **Optimizing topic coherence**: Adjust the number of topics and hyperparameters (alpha and beta) to maximize topic coherence, ensuring that the topics are meaningful and interpretable

    - **Incorporating prior knowledge**: If available, incorporate domain knowledge through priors to guide the topic discovery process

- **Bisecting K-means:**

    - **Cluster selection for splitting**: Develop or adopt criteria for selecting which clusters to split, such as those with the highest SSE or largest size, to improve the overall cluster quality

    - **Stopping criteria**: Define clear stopping criteria based on cluster stability, SSE, or a predefined number of clusters to avoid over-splitting

- **GMM:**

    - **Covariance matrix type**: Experiment with different types of covariance matrices (full, tied, diagonal, spherical) to find the best fit for the data's distribution

    - **Regularization**: Add a regularization term to the covariance matrices to ensure numerical stability, especially in high-dimensional spaces

- **PIC:**

    - **Similarity measure**: Carefully choose or design the similarity measure used to construct the graph, ensuring it captures the actual relationships in the data

    - **Graph sparsification**: Consider sparsifying the graph to retain only meaningful edges, which can reduce computational complexity and focus the clustering on strong data relationships

- **Post-clustering strategies:**

    - **Cluster validation**: Use internal and external validation indices to assess the quality of the clustering and guide improvements

    - **Result interpretation**: Engage domain experts in interpreting the clusters, which can provide insights into whether the model is capturing meaningful groupings

    - **Iterative refinement**: Clustering can be an iterative process where initial results inform subsequent rounds of feature selection, algorithm tuning, and even data collection

Improving clustering models is an iterative and exploratory process that often involves balancing the trade-offs between model complexity, interpretability, and computational efficiency. Tailoring these strategies to the specific characteristics of the dataset and the requirements of the task can lead to more meaningful and useful clustering outcomes.

Improving the performance of a classification model involves a combination of techniques applied at different stages of the model development process. Here are some strategies to enhance your model's accuracy and effectiveness:

- **Data quality and quantity:**

    - **Increase data size**: More data can help the model learn better and generalize well to unseen data. Consider augmenting your dataset if possible.

- **Data cleaning**: Remove outliers, handle missing values, and correct errors in your dataset to improve model accuracy.

- **Feature engineering**: Create new features from existing ones through domain knowledge. Well-designed features can significantly improve model performance.

- **Feature selection and dimensionality reduction**:

  - **Feature selection**: Identify and retain the most informative features while removing irrelevant or redundant ones to reduce overfitting and improve model performance.

  - **Dimensionality reduction**: Techniques such as PCA can reduce the feature space, potentially improving model efficiency and performance by removing noise.

- **Choosing the right model**:

  - **Experiment with different algorithms**: Different models have different strengths and weaknesses. Experiment with various algorithms (such as decision trees, SVMs, and ensemble methods) to find the best fit for your data.

  - **Ensemble methods**: Combining predictions from multiple models can often produce better results than any single model. Techniques such as bagging, boosting, and stacking are powerful ensemble methods to improve classification performance.

- **Hyperparameter tuning**:

  - **Grid search and random search**: These are systematic ways to search through multiple combinations of parameter values, finding the optimal settings for your model.

  - **Cross-validation**: Use cross-validation to ensure that your model's performance is robust across different subsets of your dataset.

- **Handling imbalanced data**:

  - **Resampling techniques**: Over-sample the minority class or under-sample the majority class to balance the dataset and improve model performance on minority classes.

  - **Cost-sensitive learning**: Modify the algorithm to penalize misclassifications of the minority class more than the majority class.

  - **Use appropriate metrics**: Choose evaluation metrics that provide insights into model performance on imbalanced datasets, such as F1-score, Precision-Recall AUC, or Matthews Correlation Coefficient.

- **Advanced feature engineering**:

  - **Interaction terms**: Consider adding features that capture interactions between other features if you suspect such interactions could be predictive.

  - **Polynomial features**: Use polynomial features to model non-linear relationships.

- **Model regularization**:

  - **Apply regularization techniques**: Techniques such as L1 (Lasso) and L2 (Ridge) regularization can prevent overfitting by penalizing large coefficients in the model.

- **Data transformation**:

  - **Feature scaling**: Standardize or normalize your features so that they're on the same scale. This is particularly important for models sensitive to the scale of the data, such as SVM or KNN.

  - **Data encoding**: Properly encode categorical variables using techniques such as one-hot encoding, target encoding, or embeddings, especially for algorithms that require numerical input.

- **Model updating**:

  - **Incremental learning**: Continuously update your model with new data to adapt to changes over time.

  - **Feedback loops**: Incorporate model predictions back into training to correct errors and adapt to new patterns.

- **Domain-specific techniques**:

  - **Utilize domain knowledge**: Incorporating expert knowledge can guide feature engineering, model selection, and interpretation of the results.

  - **Custom loss functions**: Design loss functions that specifically address the business problem's nuances.

- **Experimental rigor**:

  - **Systematic experimentation**: Keep a rigorous log of experiments, including model configurations, results, and observations. Tools and platforms for experiment tracking can be highly beneficial here.

  - **Peer review and collaboration**: Collaborate with peers for code reviews and discussions. Fresh perspectives can offer new insights and ideas for improvement.

Improving a classification model is an iterative process that involves experimenting with different strategies, continuously monitoring performance, and adapting to new data and insights. Balancing the trade-offs between model complexity, interpretability, and performance is the key to developing effective and reliable classification models.

Summary

In this chapter, we learned that clustering as a form of unsupervised learning offers powerful tools for discovering patterns and structures within raw, unlabeled data. By grouping similar data points together, clustering techniques such as K-means, LDA, bisecting K-means, GMM, and PIC reveal hidden insights that are crucial across various fields, from market segmentation to bioinformatics. We further covered this in this chapter. Additionally, we learned that each algorithm has its unique strengths and challenges, whether that be handling high-dimensional data, determining the optimal number of clusters, or managing the complexity of the model. Understanding and applying the right evaluation metrics is essential to assess the quality of the clustering and improve model performance, as we learned in this chapter.

As data complexity grows, the importance of robust clustering techniques in extracting valuable information and informing decisions cannot be overstated. With continuous advancements and careful application, clustering remains an indispensable part of the data scientist's toolkit, which we learned in this chapter.

In the next chapter, we will look into building recommendation systems.

7

Building a
Recommendation System

Recommendation systems are a crucial aspect of driving user engagement, conversion rates, and customer satisfaction. It is challenging to imagine services such as Netflix without personalized movie suggestions or Amazon without *"customers who bought this also liked..."* recommendations. Personalized recommendations are the key to user retention and revenue growth. E-commerce platforms such as eBay and Alibaba benefit from personalized product recommendations, and music streaming services such as Spotify curate playlists based on user preferences. Recommendation systems are more than just algorithms; they are strategic tools that enhance user satisfaction, generate revenue, and shape the digital landscape. Businesses that leverage recommendation systems power from user activities and insights gain a competitive edge.

In this chapter, we're going to cover the following main topics:

- An overview of recommendation systems
- The need for a recommendation system
- The working mechanism of recommendation systems
- The key problems and challenges in recommendation systems
- Improving the quality of recommendations
- Building a recommendation system using Apache Spark

By the end of this chapter, you will have a solid understanding and the necessary skills to build a recommendation system using Apache Spark.

Technical requirements

You can find the code files for this chapter on GitHub at https://github.com/PacktPublishing/Apache-Spark-for-Machine-Learning/tree/main/Chapter07.

An overview of recommendation systems

A recommendation system is a smart algorithm that offers personalized suggestions to users. The main goal of this system is to help users find relevant items from a vast pool of options. As shown in the following diagram, these systems use data about user preferences, behavior, and past interactions to generate individualized recommendations:

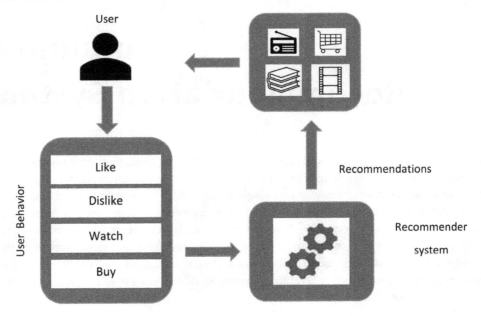

Figure 7.1 – An overview of a recommendation system

Whether they recommend movies on streaming platforms, products on e-commerce websites, or music playlists, recommendation systems improve user satisfaction and engagement. They play a critical role in our data-driven world by helping users discover content amid overwhelming choices available online.

Understanding the purpose and importance of recommendation systems

Recommendation systems serve as intelligent guides in our digital journeys, suggesting relevant items to users based on their preferences and behavior. Their purpose extends across various domains:

- **Enhanced user experience and engagement**:

  - **Purpose**: Recommendation systems analyze user behavior, preferences, and historical interactions to suggest relevant products, content, or information tailored to each individual.

  - **Impact**: By providing customized recommendations, businesses can boost customer engagement and build loyalty. A seamless experience keeps users engaged and satisfied.

- **Examples**:

  - **Netflix**: Personalized movie and TV show suggestions keep viewers hooked

  - **Spotify**: Curated playlists cater to individual music tastes

  - **YouTube**: Video recommendations guide users through a vast library of content

- **Increased sales and revenue**:

  - **Purpose**: Recommendations drive revenue. When users discover relevant products quickly, they're more likely to make a purchase.

  - **Impact**: Recommendations drive conversions. Users are more likely to make a purchase when guided toward relevant items.

  - **Example**: Amazon attributes 35% of its revenue to recommendation algorithms.

- **Reduced information overload**:

  - **Purpose**: In a world of endless choices, recommendation systems filter noise

  - **Impact**: Users appreciate platforms that simplify decision-making, leading to higher satisfaction and repeat visits

  - **Examples**:

    - **Google News**: By curating news articles based on user interests and behavior, Google News streamlines information consumption.

    - **eBay**: Providing personalized product recommendations can greatly improve the shopping experience, particularly for categories with a wide range of inventory.

    - **PubMed**: The recommendation system suggests relevant research papers to medical professionals, providing tailored suggestions based on their specialty and interests, instead of requiring them to manually search through thousands of articles.

- **Business metrics and brand royalty**:

  - **Purpose**: Recommendations to impact **key performance indicators** (**KPIs**) and customer loyalty.

  - **Impact**:

    - Click-through rates, conversion rates, and average order value benefit from personalized suggestions

    - Users who spend more time on platforms due to effective recommendations become loyal customers

  - **Examples**: Amazon's recommendation engine significantly impacts revenue. By suggesting related products (cross-sell) and complementary items (upsell), they increase average order value. Personalized recommendations keep users engaged, encouraging repeat purchases and fostering loyalty.

In today's competitive landscape, understanding recommendation systems is essential for data scientists fine-tuning algorithms or entrepreneurs building start-ups. Such systems are essential for enhancing user satisfaction, driving revenue, and shaping digital interactions.

An overview of various recommendation approaches

Recommendation systems are essential tools for enhancing user experiences by suggesting relevant content, products, or services. These systems leverage different techniques to provide personalized recommendations. Let's explore some common approaches.

Collaborative filtering methods

Collaborative filtering is a popular recommendation technique that leverages user-item interactions to make personalized recommendations. Here are some features of this method:

- Collaborative filtering is a recommendation method used to recommend items to users based on the similarity of the preferences and behaviors of the other users with those items

- The core idea is to group users based on their behavior and use general group characteristics to suggest relevant items to a target user

- It operates on the principle that similar users (in terms of behavior) share similar interests and tastes

The following diagram captures the mechanism of collaborative filtering:

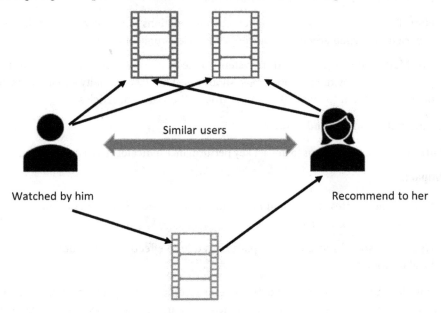

Figure 7.2 – A working mechanism of collaborative filtering

Different types of collaborating filtering methods are as follows:

- **Memory-based collaborative filtering**: Memory-based collaborative filtering is a type of method that relies solely on past user-item interactions. You can visualize this as a matrix where the rows represent users, the columns represent items, and each cell contains interaction data such as ratings. There are two approaches to this method – user-user and item-item approaches. Both of these approaches compute similarities between users or items based on these interactions. Using these similarities, items are recommended to users. However, it's important to note that these methods don't build explicit models.

- **Model-based collaborative filtering**: Model-based approaches to recommendation systems differ from memory-based methods in that they create predictive models utilizing interaction data. Techniques such as matrix factorization and latent factor models are examples of this approach. These models capture underlying patterns and relationships within the data, resulting in more accurate recommendations. However, these models require more computational resources and are susceptible to the cold start problem, which arises when new users or items lack sufficient data.

Content-based methods

Content-based filtering is a recommendation system that suggests items based on their attributes, rather than a user's interactions. For example, if a user enjoys watching action movies, the system will recommend similar action movies, as shown in the following diagram. This method focuses on item features such as genres, keywords, or descriptions. Content-based methods work well when there's a lot of item metadata available, but they struggle with suggesting diverse items beyond user preferences, also known as serendipity.

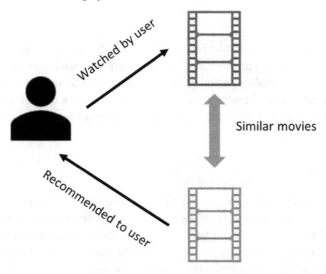

Figure 7.3 – A working mechanism of a content-based recommendation

Hybrid recommendation systems

Hybrid recommendation systems are designed to overcome the limitations of individual methods by combining collaborative filtering and content-based approaches. These systems leverage both user-item interactions and item attributes to provide robust recommendations that balance personalization and diversity. Weighted hybrid models and feature combination approaches are some examples of hybrid systems that can be used for this purpose.

Deep learning-based recommendations

The field of recommendation systems has been greatly influenced by recent advancements in deep learning. With the ability to learn complex patterns from large-scale data, neural networks have proven to be a valuable tool. Techniques such as **neural collaborative filtering** (**NCF**) and **recurrent neural networks** (**RNNs**) have been particularly effective in enhancing the quality of recommendations. However, it's important to note that these methods require a significant amount of data and computational resources to be effective.

In summary, recommendation systems are important for providing personalized experiences in various domains. They silently guide your choices, making your digital journey more enjoyable and efficient, whether you are binge-watching shows on Netflix or shopping online.

The need for a recommendation system

The need and value of recommendation systems can be understood through various perspectives such as personalization, user engagement, business growth, data utilization, content discovery, and bridging supply and demand. We will explore them further in the following subsections.

Personalization

In today's world where we are exposed to a lot of information, personalization is not just a nice-to-have feature but also a necessity. Recommendation systems use advanced algorithms to analyze large amounts of data in order to provide a personalized experience to each user. By customizing content to suit individual preferences and tastes, these systems help create a unique and satisfying user experience. Personalization promotes a deeper connection between users and platforms, resulting in more meaningful and relevant interactions.

User engagement

Recommendation systems play a key role in improving user engagement. By showcasing options that are aligned with users' preferences and previous activities, these systems boost the chances of user interaction. This could mean watching another video on a streaming service, reading another article on a news website, or buying more products from an online store. Improved engagement not only enhances user satisfaction but also results in longer session times and more frequent visits, which in turn strengthens user loyalty.

Business growth

Recommendation systems are highly valuable from a business perspective. They have the potential to increase sales and content consumption by offering accurate and timely suggestions, which leads to higher conversion rates and increased average order values. For example, Amazon's recommendation engine is responsible for a significant portion of its sales by suggesting products based on browsing history, purchase history, and what other customers have viewed or purchased. This targeted approach not only boosts sales but also enhances the efficiency of marketing efforts, by aligning product placements with consumer inclinations.

Data utilization

The data generated by users daily is a valuable resource for businesses, and recommendation systems are the key to unlocking its full potential. These systems use user data, such as transaction history, browsing habits, ratings, and social interactions, to develop advanced models that predict user preferences. This not only helps in optimizing inventory and content management based on demand forecasts but also in identifying emerging trends and user segments, thereby informing product development and strategic decisions.

Content discovery

One of the biggest benefits of recommendation systems for users is content discovery. With an overwhelming range of choices in products, services, and content, users can easily get confused or stick to their familiar choices. Recommendation systems help users discover new and exciting content that they might not have found on their own, expanding their horizons and enhancing their overall experience. This is particularly evident in platforms such as Netflix or Spotify, where the discovery of new movies, shows, or music is a crucial part of the user experience.

Bridging supply and demand

Recommendation systems play an essential role in connecting users with the right products or content from a provider's inventory. This helps to ensure that valuable products or content reach their intended audience, and it also maximizes the utilization of available resources, which contributes to a more efficient market.

Recommendation systems are an essential part of the digital world today. They play a crucial role in user experience by driving personalization, engagement, and business growth. These systems use data to understand and predict user preferences, which helps businesses meet user needs more effectively. They also facilitate content discovery and enhance overall satisfaction. As technology advances, the significance and complexity of recommendation systems are likely to increase, highlighting their pivotal role in shaping the digital landscape.

The working mechanism of recommendation systems

Let's understand the working mechanism of recommendation systems, such as their workflow, pros, and cons, and applications for each of the recommendation system types.

Content-based recommendation systems

Content-based recommendation systems are a popular way to offer personalized recommendations to users. These systems work on the principle that a user's interests can be predicted based on their interactions with items, such as viewing history or purchasing behavior. The main objective of this approach is to suggest items that are similar to those that the user has previously shown interest in.

The workflow of a content-based recommendation system

Here's how this system works:

1. **Feature extraction and preprocessing**:

   - **Feature extraction**: We start by identifying the essential features for recommendations. These features describe items such as movies, articles, and products, and they can include genres, keywords, tags, or textual descriptions.

   - **Preprocessing**: Clean and preprocess data by tokenizing text, removing stop words, and converting features to a suitable format.

2. **Creating user profiles**:

   - Each user is represented by a profile that captures their preferences. This profile is built using the features extracted from the items they have interacted with.

   - If a user watches sci-fi movies frequently, their profile will highlight the sci-fi genre.

3. **Extracting user preferences**:

   - Based on user interactions (for example, ratings, clicks, and views), extract relevant features from the items they have engaged with

   - These preferences contribute to building the user profile

4. **Learning user profiles via matrix algebra**:

   - Use matrix algebra techniques (such as matrix factorization) to learn the user profiles

   - The goal is to map user preferences to latent factors that capture their tastes

5. **Building recommendations**:

   - Based on a user's profile, provide suggestions for items that match their preferences

   - Calculate the similarity between the user profile and other items using various similarity measures (for example, cosine similarity and Pearson correlation)

   - Recommend items with high similarity scores

6. **Deploying the recommendation system**:

   - Integrate the recommendation system into the application or platform

   - Continuously update user profiles as new interactions occur

   - Monitor system performance and gather feedback for further improvements

The following figure captures the various elements, steps, and processes involved in a content-based recommendation system:

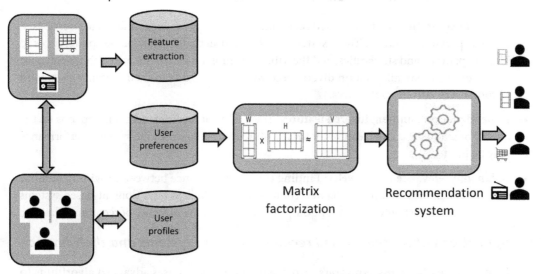

Figure 7.4 – An overview of a content-based recommendation system

The pros of content-based recommendation systems

- **User-specific recommendations**:

  - Content-based systems personalize recommendations for individual users, without relying on data from other users, making them scalable for large user bases.

  - These systems are designed to identify the specific interests of a user and recommend niche items that align with their tastes. This personalized approach aims to provide users with a more tailored and enjoyable experience.

- **Domain knowledge utilization**:

  - Content-based systems personalize recommendations using hand-engineered features such as genres, keywords, and descriptions

  - When there is domain knowledge available, they can excel by creating precise models of user preferences

- **Avoiding the "new item problem"**:

  - Content-based approaches are more effective than collaborative filtering in handling new items. This means that they can provide accurate recommendations for new items that have not been previously rated or reviewed.

  - Content-based systems can recommend items based on their descriptions without relying on historical interactions.

The cons of content-based recommendation systems

- **Limited exploration**: Content-based recommendation systems suggest items to users based on their previous interests. These systems analyze a user's past behavior and preferences to identify patterns and similarities, and then they recommend items that match those patterns. However, they may miss out on diverse recommendations due to their inability to expand beyond users' current preferences.

- **Dependency on hand-engineered features**: The quality of recommendations depends on the quality of manually crafted features. If these features are suboptimal, the system's performance may be limited.

- **Balancing relevance and novelty**: Finding the perfect balance between providing familiar recommendations and introducing new content can be a difficult task. Content-based systems can at times suggest only familiar options, which can limit the diversity of recommendations.

The applications of content-based recommendation systems and their impact

- **E-commerce platforms**: Amazon's recommendation system uses advanced algorithms to analyze product metadata, which includes genres, descriptions, and customer reviews. Based on a user's browsing history, past purchases, and explicit ratings, the system suggests related products. For instance, if you buy a sci-fi book, the system will recommend similar titles or related merchandise that you might be interested in.

- **Streaming services**: Netflix's recommendation system utilizes both content-based and collaborative filtering. The system analyzes metadata such as genre, cast, and director to suggest similar content after users watch a movie or show. For example, if someone enjoys action movies, then the system recommends other action-packed titles.

- **Music platforms**: Spotify uses song features such as tempo, genre, and artist to create personalized playlists. For instance, if you listen to upbeat pop songs, it curates playlists with similar tracks. This enhances user satisfaction and engagement.

- **News aggregators and social media**: Facebook's recommendation algorithm analyzes user interactions, such as likes, shares, and comments, as well as content attributes such as topics and keywords to suggest relevant posts, articles, and friends to keep users engaged.

- **Video sharing platforms**: YouTube suggests videos based on content similarity. For example, if you watch cooking tutorials, it recommends more cooking-related content. The system adapts to user preferences over time, maximizing satisfaction.

- **Online advertising**: Google Ads uses content-based recommendations to display relevant ads by analyzing website content, user behavior, and keywords to target specific audiences. For instance, searching for hiking gear will show ads related to outdoor equipment.

In summary, content-based recommendation systems improve user experiences, drive sales, and streamline content discovery across various domains.

Collaborative filtering recommendation systems

Collaborative filtering is a technique that recommends items to users based on the preferences and behaviors of other users who share similar tastes. Unlike traditional recommendation methods that rely on item features, collaborative filtering analyzes interactions among users and their preferences to generate recommendations. The method is based on the concept that similar users tend to like similar things and aims to capture those similarities, providing more accurate recommendations to users.

The workflow of collaborative filtering in recommendation systems

This method works as follows:

1. **Data collection and user behavior**:

    I.   Develop a comprehensive user behavior database using data from a large number of users.

    II. Collect data such as ratings, clicks, views, or explicit preferences (for example, movie ratings and product purchases).

2. **User-item interaction matrix**:

    I.   Construct a matrix where each row represents a user, and each column represents an item (for example, movies and products).

    II. Populate the matrix with user feedback – explicit (numerical ratings) or implicit (watched, clicked, and liked).

3. **Similarity calculation:**

    I.   Measure similarity between users or items.

4. **User-based collaborative filtering:**

    I.   Identify users with similar preferences.

    II. Recommend items liked by users similar to the target user. For instance, if user A and user B have similar movie preferences, recommend movies liked by user B to user A.

5. **Item-based collaborative filtering:**

    I.   Focus on similarities between items.

    II. Recommend items similar to those that the user has already engaged with. If user A enjoyed *The Matrix*, suggest other sci-fi movies

6. **Embeddings and automatic learning:**

    I.   Assign scalar values (embeddings) to movies and users.

    II.   These embeddings capture characteristics (for example, child versus adult content).

    III. The product of movie and user embeddings predicts user preferences.

    IV.   Learn embeddings automatically without manual feature engineering.

7. **Deployment and recommendations:**

    I.   Integrate the collaborative filtering system into an application or platform.

    II.   Continuously update user profiles and monitor system performance.

    III. Provide personalized recommendations to users based on similarity and embeddings.

Let's look at the pros and cons of collaborative filtering.

The pros of collaborative filtering

- **Improved recommendation accuracy:**

  - Collaborative filtering leverages user behavior patterns to make accurate recommendations. By analyzing past interactions, it identifies similar users or items, leading to better predictions.

- **Robustness and adaptability:**

  - Collaborative filtering adapts well to changes in user preferences and item availability

  - It doesn't rely on explicit domain knowledge or feature engineering, making it versatile across different domains

- **Mitigates the cold start problem**:

  - The cold start problem occurs when new users or items have a limited interaction history

  - Collaborative filtering can still provide recommendations by relying on similar users or items, even when direct data is scarce

- **Enhances user satisfaction**:

  - By suggesting relevant items based on user behavior, collaborative filtering improves user engagement and satisfaction

  - Users appreciate personalized recommendations that align with their interests

The cons of collaborative filtering

- **Complex implementation and maintenance**:

  - Collaborative filtering requires careful implementation and tuning

  - Managing large-scale recommendation systems can be challenging due to computational demands and scalability issues

- **Parameter tuning and weights**:

  - Fine-tuning parameters (for example, similarity metrics and neighborhood size) is essential for optimal performance

  - Incorrect parameter choices can lead to suboptimal recommendations

- **Potential biases**:

  - Collaborative filtering can introduce biases due to the inherent nature of user-item interactions

  - Biases may arise from popularity bias (favoring popular items) or user demographics

- **Limited contextual information**:

  - Collaborative filtering primarily relies on past behaviors and lacks context (for example, time, location, and user demographics)

  - Incorporating additional features (side information) can be challenging within the collaborative filtering framework

The following figure captures the various elements, steps, and processes involved in a collaborative filtering recommendation system:

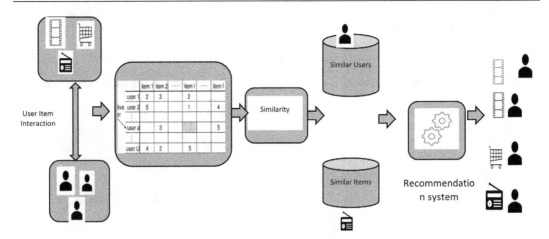

Figure 7.5 – An overview of collaborative filtering

In summary, collaborative filtering uses user interactions and similarities to provide personalized recommendations, improving user experiences in various domains.

The workflow of user-based collaborative filtering

Personalized recommendations are made by leveraging the preferences and behaviors of similar users. This technique is used to recommend items to users based on their interests and actions.

The workflow is as follows:

1. **User similarity calculation**:

    I.   First, we create a matrix where each row represents a user and each column represents an item (for example, a movie, product, or song).

    II. We calculate the similarity between users based on their interactions with items. Common similarity metrics include cosine similarity, Pearson correlation, or the Jaccard index.

    III. Users who have similar preferences (i.e., high similarity scores) are considered "neighbors."

2. **Recommendation generation**:

    I.   For a target user (the one we want to recommend items to), we identify their neighbors (similar users).

    II. We look at the items the neighbors have liked, rated, or interacted with.

    III. Items that the neighbors enjoyed but the target user hasn't seen yet become potential recommendations.

    IV. The system suggests these items to the target user.

In summary, user-based collaborative filtering helps personalize recommendations by tapping into the collective wisdom of users with similar tastes.

Real-world examples of user-based collaborative filtering

- **Movie recommendations**:

  - Imagine two users, user A and user B, who have similar movie preferences.

  - If user A has watched and liked movie X, the system will recommend movie X to user B.

  This approach is widely used by streaming platforms such as Netflix to suggest movies based on what similar users enjoy.

- **E-commerce product recommendations**:

  - Suppose a group of users has purchased products similar to those in a user's shopping history

  - If these users also bought a specific item, the system will recommend that item to the user

  - E-commerce platforms such as Amazon utilize this technique to personalize product recommendations

- **Music recommendations**:

  - When users have similar music preferences, collaborative filtering can suggest songs they might enjoy

  - If user C listens to artists similar to those favored by user D, the system recommends relevant music to user D

  Music streaming services such as Spotify employ collaborative filtering to enhance user playlists.

Item-based collaborative filtering

This is a recommendation technique that focuses on the similarity between items rather than users. Let's explore how it works.

The workflow of item-based collaborative filtering in recommendation systems

Here's how it works:

1. **Item similarity calculation**:

   I.   We create a matrix where each row represents an item (for example, a movie, product, or song), and each column represents a user.

   II. The similarity between items is calculated based on user interactions. Common similarity metrics include cosine similarity, the Jaccard index, or adjusted cosine similarity.

   III. Items that tend to be liked or interacted with by the same users are considered similar.

2. **Recommendation generation**:

    I.   For a target user, we look at the items they have already rated or interacted with.

    II.   We identify similar items based on their interactions with other users.

    III. Items that are similar to those that the user has already liked become potential recommendations.

    IV. The system suggests these similar items to the user.

The pros of item-based collaborative filtering

- Item-based filtering is computationally efficient because it doesn't require calculating user similarities
- It handles the "cold start" problem better for new users, since it relies on item-item relationships

The cons of item-based collaborative filtering

- Item-based filtering may miss out on personalized recommendations if users have unique preferences
- It requires a large dataset of user interactions for accurate recommendations

In summary, item-based collaborative filtering helps users discover related items based on their existing preferences.

Real-world examples of item-based collaborative filtering

- **Movie recommendations**:
  - Imagine a movie streaming platform where users rate and interact with movies
  - Item-based collaborative filtering identifies movies that are similar to those the user has already liked
  - If a user enjoyed movie X, the system recommends other movies highly similar to it

  This approach helps personalize movie suggestions based on item similarity, enhancing the user's viewing experience.

- **E-commerce product recommendations**:
  - Online retailers such as Amazon use item-based filtering to recommend products
  - If a user has purchased or viewed a specific product, the system identifies related items

  For instance, if a user bought a camera, the system might suggest camera accessories (lenses or tripods) based on item similarity.

  This technique enhances cross-selling and helps users discover relevant products.

- **Music playlists**:

  - Music streaming services employ item-based collaborative filtering to create personalized playlists.

  - If a user enjoys songs by a particular artist or genre, the system recommends similar tracks.

  For example, if a user likes songs by artist A, the system suggests other songs by artist A or similar artists.

  By leveraging item similarity, music platforms enhance user engagement and discovery.

Now, let's look at one of the algorithms in Apache Spark.

Alternating Least Squares (ALS) – the collaborative filtering algorithm in Apache Spark

ALS is a matrix factorization technique commonly used for collaborative filtering in recommendation systems. It aims to predict missing values (such as user-item ratings) by decomposing a large matrix into two lower-dimensional matrices. ALS falls under the collaborative filtering category.

The benefits of ALS include the following:

- **Scalability**: ALS works well with large datasets and parallel processing (ideal for Spark)
- **Implicit feedback**: It handles implicit feedback (for example., clicks and views) effectively
- **Cold start**: It handles new users or items with limited interaction history

ALS can be applied in e-commerce as follows:

- In e-commerce, ALS can recommend products to users based on their historical interactions (purchases, clicks, and so on)
- By learning latent features, ALS predicts missing ratings and suggests personalized products
- Research shows that ALS significantly reduces the **Root Mean Squared Error** (**RMSE**), improving recommendation accuracy

Imagine we have a big table with users and their ratings for different movies (or products). Some users have rated movies they liked, but many cells in the table are empty because users haven't seen or rated all movies.

The goal is as follows:

- ALS should help us fill in those empty cells by making predictions
- We want to find two special matrices – one for users (let's call it U) and one for movies (let's call it P)
- When we multiply these matrices, we should get an estimate of the original user-movie ratings

The workflow of ALS

The ALS algorithm works as follows. ALS takes turns improving U and P.

1. First, it fixes U and adjusts P to minimize the prediction error.
2. Then, it fixes P and adjusts U to make better predictions.
3. It keeps alternating until it finds good estimates for both U and P.

Let's see this in action with the help of an example. Suppose we want to predict how user 1 would rate movie B:

1. We start with random values in U and P (such as guessing).
2. ALS adjusts these values based on existing ratings.
3. For instance, if user 1 liked movie A (and rated it 5), we adjust U and P to match this.
4. We repeat this process, refining our estimates until they converge.

The pros of using "least squares" for convergence of the ALS model

- ALS aims to minimize the difference between predicted and actual ratings
- It's like fitting puzzle pieces together to get the best match
- The "least squares" part means finding the best fit by minimizing errors

The key problems and challenges in recommendation systems

When working with recommendation systems, we can encounter two types of problems:

- **Cold start problem**: The cold start problem occurs with new users or items that lack sufficient data for profiling
- **Data sparsity**: Data sparsity arises when there are many users and items but only a small subset interacts

Both challenges impact recommendation accuracy. Let's understand this more by exploring them in detail in the context of different recommendation methods.

Cold start

Cold start in recommendation systems refers to the challenge when a new system lacks sufficient interaction history for users or items. Without enough data, making accurate predictions becomes difficult, impacting recommendation quality and efficiency. The following are some of the techniques to overcome the challenges associated with the cold-start problem:

- **Content-based recommendations**:

  - For new items (cold-start items), leverage their content features

  - Analyze item attributes (such as genre, keywords, and descriptions) to recommend similar items

  - Content-based methods work well when item characteristics are well-defined

- **Popularity-based recommendations**:

  - Initially, recommend popular items to new users

  - Popularity-based recommendations serve as a baseline until personalized data is available

  For example, suggest trending movies or bestsellers

- **Hybrid approaches**:

  - Combine content-based and collaborative filtering techniques

  - Use content features for cold-start items and collaborative filtering for existing items

  - Hybrid models balance accuracy and diversity

- **Demographic information**:

  - Collect user demographic data (age, gender, and location)

  - Use this demographic information to make initial recommendations

  - As users interact more, switch to personalized recommendations

- **Exploration versus exploitation**:

  - Employ exploration strategies for new users/items

  - Show diverse recommendations to learn user preference

  - Gradually shift toward exploitation (personalized recommendations)

- **Implicit feedback**:
  - Use implicit feedback (clicks and views) for cold-start users
  - Analyze patterns in implicit interactions to infer preferences
  - Implicit feedback helps even when explicit ratings are sparse

- **Active learning**:
  - Prompt users to provide feedback during the cold-start phase
  - Ask them to rate or indicate preferences for a few items
  - Use this feedback to bootstrap the recommendation process

- **Contextual information**:
  - Consider contextual factors (time, location, and device)
  - Context-aware recommendations adapt to user situations
  - For example, suggest nearby restaurants based on location

- **Warm-up period**:
  - Delay personalized recommendations for new users
  - Allow a warm-up period where users explore the system
  - Gradually transition to personalized suggestions

- **User onboarding**:
  - Guide new users through the system
  - Explain how recommendations work and encourage interactions
  - Provide tooltips or tutorials

Data sparsity

Data sparsity occurs when transactional or feedback data is sparse, making it challenging to identify neighbors. For example, a new product will typically have very few user reviews. This limitation impacts the quality of recommendations and the effectiveness of collaborative filtering in recommendation systems. The following are some of the techniques to overcome the challenges associated with the data sparsity problem:

- **Content-based recommendations**:
  - Leverage item content features (for example, genre, keywords, and descriptions)
  - Recommend items similar to those with known content features
  - Content-based methods work well for new or sparsely rated items

- **Popularity-based recommendations**:

  - Initially, recommend popular items to all users

  - Popularity-based suggestions serve as a baseline

  - Gradually transition to personalized recommendations as more data accumulates

- **Hybrid approaches**:

  - Combine collaborative filtering (user-item interactions) with content-based methods

  - Use collaborative filtering for existing users/items and content-based filtering for new ones

  - Hybrid models balance accuracy and diversity

- **Matrix factorization techniques**:

  - Techniques such as ALS factorize the user-item interaction matrix

  - These methods handle sparsity by learning latent factors from available data

  - Regularization helps prevent overfitting

- **Implicit feedback handling**:

  - Implicit feedback (clicks and views) is valuable for sparse data

  - Weight interactions based on confidence (for example, more clicks imply stronger preference)

  - Use implicit feedback to infer user preferences

- **User and item embeddings**:

  - Learn dense representations (embeddings) for users and items

  - Embeddings capture latent features and handle sparsity

  - Neural network-based models (for example, Word2Vec and item2vec) learn embeddings effectively

- **Side information**:

  - Incorporate additional data sources (that is, side information)

  - User demographics, social connections, or context can enhance recommendations

  - For example, consider user location or time of interaction

- **Active learning:**

  - Encourage users to provide feedback

  - Prompt them to rate items or indicate preferences

  - Use this feedback to improve recommendations

- **Temporal dynamics:**

  - Consider how user preferences change over time

  - Decay older interactions to give more weight to recent ones

  - Time-aware models adapt to evolving preferences

- **Transfer learning:**

  - Transfer knowledge from related domains

  - If data is sparse in one domain, leverage information from another domain

  - Cross-domain recommendations enhance coverage

Now that we understand the problems and challenges associated with the recommendations, let's explore different ways of improving recommendations.

Improving the quality of recommendations

Like any other machine learning algorithm, the recommendation algorithms also suffer performance issues. However, there are several techniques to improve the quality of the recommendations. Let's look at each of those techniques:

- **Hyperparameter tuning:**

  - Experiment with different hyperparameters for the ALS model

  - Key hyperparameters include `rank` (the number of latent factors), `maxIter` (the number of iterations), and `regParam` (the regularization parameter)

  - Use techniques such as grid search or random search to find optimal values

- **Feature engineering:**

  - Incorporate additional features beyond user-item interactions (for example, consider user demographics, time of interaction, or contextual information)

  - Feature engineering can improve model performance significantly

- **Implicit feedback handling**:

  - ALS is commonly used for implicit feedback data (for example, clicks and views)

  - Adjust the model to handle implicit feedback effectively

  - Explore techniques such as weighting interactions or using confidence scores

- **Cold start strategies**:

  - Address the cold start problem for new users or items

  - Use content-based recommendations or hybrid approaches until sufficient interactions are available

- **Ensemble methods**:

  - Combine ALS with other recommendation algorithms

  - Ensemble methods (for example, stacking) can enhance overall performance

- **Regularization and overfitting**:

  - Regularize a model to prevent overfitting

  - Tune the regularization parameter (`regParam`) carefully

- **Model evaluation**:

  - Use appropriate evaluation metrics (for example, RMSE and MPR) to assess model performance

  - Compare ALS variants and choose the best-performing one

Now, it's time to look at assessing the quality of recommendations.

Evaluating the recommendations

Ranking algorithms, also known as recommendation systems, are designed to provide users with a relevant set of items or documents based on certain training data. The definition of relevance may vary, depending on the application. Metrics for ranking systems aim to measure the effectiveness of these recommendations in different contexts. Some metrics compare the recommended documents to a predetermined set of relevant documents, while other metrics can use numerical ratings.

The following table lists some of the common metrics used to evaluate a recommendation:

Metric	Definition
Precision at k	Precision at k is a performance metric used to evaluate the accuracy of a recommendation system. It measures the proportion of the first k recommended documents that are truly relevant to the user, averaged across all users. Note that the order of the recommendations is not considered in this metric.
Mean average precision (MAP)	MAP is a measure of how many of the recommended documents are in the set of true relevant documents, where the order of the recommendations is taken into account (that is, the penalty for highly relevant documents is higher).
Normalized discounted cumulative gain (NDCG)	NDCG at k is a metric used to measure the relevance of recommended documents in comparison to true relevant documents. It calculates the percentage of top k recommended documents that are relevant, taking into account the order of recommendations. This means that documents are assumed to be in decreasing order of relevance. Unlike precision at k, NDCG at k considers the order of recommendations while evaluating the quality of recommendations.

Table 7.1 – An overview of the metrics used to evaluate a recommendation system

After learning how to evaluate a recommendation, we will now build a recommendation system.

Building a recommendation system using Apache Spark

Let's walk through a code example to build a recommendation system.

The following code snippet imports all the required libraries and creates a Spark session:

```
from pyspark.sql import SparkSession
from pyspark.ml.recommendation import ALS
from pyspark.ml.evaluation import RegressionEvaluator
# Create a Spark session
spark = SparkSession.builder\
                .appName("AmazonProductRecommendation")\
                .getOrCreate()
spark.sparkContext.setLogLevel("ERROR")
```

Then, we will load the data into the DataFrame and display it as an output:

```
df = spark.read.format('csv')\
        .option('header','true')\
        .option('inferSchema', 'true')\
        .option('timestamp', 'true')\
```

```
            .load('s3a://test234/ratings.csv')
# Prepare the data (assuming you have columns: 'user', 'item',
'rating')
df = df.withColumnRenamed('userId', 'user')\
        .withColumnRenamed('movieId', 'item')\
        .withColumnRenamed('rating', 'rating')
df.limit(6).toPandas()
```

Here's the output:

	userId	movieId	rating	timestamp
0	1	1	4.0	964982703
1	1	3	4.0	964981247
2	1	6	4.0	964982224
3	1	47	5.0	964983815
4	1	50	5.0	964982931
5	1	70	3.0	964982400

Figure 7.6 – A sample recommendation dataset

The next code snippet creates ALS by fitting the data to the model:

```
als = ALS(
          rank=10, maxIter=10, regParam=0.1, userCol='user',
          itemCol='item', ratingCol='rating')
model = als.fit(df)
```

The following code snippet evaluates the model performance through evaluation metrics:

```
evaluator = RegressionEvaluator(metricName="rmse",
    labelCol="rating", predictionCol="prediction")
predictions = model.transform(df)
rmse = evaluator.evaluate(predictions)
print(f"Root Mean Squared Error (RMSE): {rmse}")
```

The following code snippet runs the recommendations for all the users. It displays the recommendation output:

```
# Make recommendations for users (for example, user 0)
user_recs = model.recommendForAllUsers(5)
user_0_recs = user_recs.where(user_recs.user == 24) \
```

```
    .select(
        "recommendations.item",
        "recommendations.rating"
    ).collect()
print(f"Recommendations for user 24: {user_0_recs}")
```

We will get the following results:

```
Root Mean Squared Error (RMSE): 0.5933279277739512
Recommendations for user 24: [Row(item=[33649, 3379, 171495, 74282, 7121], rating=[4.727833271026611, 4.6474504470825195, 4.5
25336265563965, 4.409307479858398, 4.388001441955566])]
```

Figure 7.7 – The recommendation evaluation metrics output

This concludes our walk-through of the code example to build a recommendation system.

Summary

In this chapter, we explored the complex world of recommendation systems, uncovering the techniques and methodologies that make personalized recommendations possible. We started with an overview of recommendation systems, emphasizing their significance in today's digital economy. From enhancing user engagement to driving revenue growth, the impact of these systems is profound across various industries, including e-commerce and streaming services.

We delved into the types of recommendation systems, distinguishing between content-based, collaborative filtering, and hybrid approaches. Each method has its unique strengths and challenges, and understanding these helps to design systems that best suit specific needs.

The chapter also addressed key problems faced by recommendation systems, such as the cold start problem, data sparsity, and scalability. These challenges are critical in ensuring the effectiveness and efficiency of recommendation systems, especially as they scale to handle vast amounts of data.

We then moved on to practical aspects, discussing how to build recommendation models using Apache Spark. By leveraging Spark's MLlib and matrix factorization techniques, we can implement collaborative filtering algorithms that are both robust and scalable. Evaluating model performance using metrics such as accuracy and precision ensures that the recommendations provided are reliable and valuable to users.

Throughout this chapter, you gained insights into the foundational concepts and advanced techniques in building recommendation systems. You now possess the skills to design, implement, and evaluate effective recommendation models that can significantly enhance the user experience and business outcomes.

In the next chapter, we will learn more about frequent pattern mining.

8

Mining Frequent Patterns

Frequent pattern mining is a crucial task in data mining. Its objective is to identify recurring associations or relationships within a dataset. These patterns may represent items, sequences, or substructures that appear together frequently. By detecting frequent patterns, we gain insights into correlated items and uncover hidden regularities. For instance, in market basket analysis, frequent patterns expose which products customers often purchase together. Understanding these patterns can help with personalized recommendations, anomaly detection, and predictive modeling. Frequent patterns play a crucial role in extracting meaningful knowledge from large datasets, guiding decision-making, and enhancing various applications across domains.

In this chapter, we're going to cover the following main topics:

- The basic concepts of frequent patterns and the significance of discovering patterns and rules
- Frequent pattern mining applications and case studies
- The key challenges in frequent pattern mining
- Frequent pattern mining algorithms
- Developing the model using scalable frequent pattern mining algorithms

By the end of this chapter, you will have a solid understanding of the necessary skills to mine frequent patterns using Apache Spark.

Technical requirements

You can find the code files for this chapter on GitHub at `https://github.com/PacktPublishing/Apache-Spark-for-Machine-Learning/tree/main/Chapter08`.

The basic concepts of frequent patterns and the significance of discovering patterns and rules

Frequent pattern mining (**FPM**) is a critical task in data mining that involves identifying patterns or itemsets that appear frequently within a dataset. These patterns are useful for discovering associations, correlations, and structures among large sets of data in various domains, such as market basket analysis, web mining, bioinformatics, and network security.

FPM is a powerful tool for data analysis, enabling the discovery of relationships hidden in large datasets that are not readily apparent. Its application spans numerous domains, helping businesses and researchers extract valuable insights from vast amounts of data.

Let's understand the visualization of market basket analysis with the help of the following diagram. Each box in each of the shopping carts represents an item purchased.

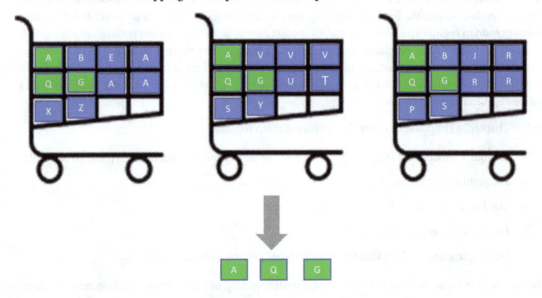

Figure 8.1 – An overview of the shopping carts, each filled with different items

In this diagram, there are three shopping carts filled with various items purchased from a grocery store. Each item is represented by a box in the cart. Upon examining all the items in the three shopping carts, it is evident that items A, Q, and G are present in all three carts. This indicates that items A, Q, and G are frequently bought together. Based on this pattern, if someone purchases items A and Q, they are recommended to also purchase item G.

The key concepts of frequent pattern mining include the following:

- **Itemsets:** An itemset is a collection of one or more items. For example, in the context of a supermarket, an itemset might be {milk, bread, butter}.

- **Support**: The support of an itemset refers to the proportion or frequency of transactions in a dataset that contains the itemset. It's a measure of how popular or common the itemset is within the given data. For example, if 100 out of 1,000 transactions contain {milk, bread}, then the support for the {milk, bread} itemset is 10%.

- **Frequent itemsets**: An itemset is considered frequent if its support is greater than or equal to a user-specified support threshold. This threshold is often set based on the specific requirements of the application.

- **Association rules**: These are implications of the form $X \Rightarrow Y$, where X and Y are itemsets. The rule means that whenever the X itemset is present, the Y itemset is likely to be present in a transaction. Two key metrics for association rules are as follows:

  - **Confidence**: A measure of the reliability of the inference made by the rule. For the rule $X \Rightarrow Y$, it is the proportion of transactions with X that also contain Y.

  - **Lift**: The lift value measures how much more often X and Y occur together than expected if they were statistically independent. A lift value greater than 1 indicates that X and Y appear more often together than expected, showing a positive relationship.

In this section, we learned about the key concepts of FPM. In the next section, we will look at the FPM algorithm.

Frequent pattern mining applications and case studies

FPM is a critical technique in data analysis that helps uncover hidden structures in large datasets. It has a widespread application across various industries, enhancing decision-making and strategy formulation. Here are the top applications of FPM:

- **Insight into customer behavior**: In retail and e-commerce, analyzing transaction data to identify frequently purchased items together helps businesses understand customer buying habits. This knowledge allows companies to optimize their marketing strategies, such as through targeted advertising, personalized recommendations, cross-selling, and upselling. For example, if bread and butter are frequently bought together, a store might place these items near each other to increase sales of both.

- **Enhancing product placement and store layout**: Discovering frequent itemsets helps retailers design better store layouts by placing items that are often bought together in close proximity. This not only enhances customer satisfaction by making shopping more convenient but also increases the chance of impulse buys, thereby boosting sales.

- **Inventory management**: By understanding which items are frequently purchased together, businesses can more effectively manage inventory levels, ensuring that popular items are well-stocked and optimally placed. This reduces instances of stockouts and overstocks, leading to more efficient operations and reduced costs.

- **Developing effective promotions and discounts**: Association rules can help businesses develop bundled offers or discounts on items that are frequently bought together. This can attract more customers and encourage them to buy more items than they originally planned. Effective promotions directly impact sales volumes and customer retention.

- **Market basket analysis**: This is perhaps the most classic application of FPM. It involves analyzing customer transaction data to find frequent patterns of items bought together. Retailers use this information for product placement, promotions, and cross-selling strategies, optimizing the layout of stores and online platforms to increase sales.

- **Recommendation systems**: FPM is fundamental in building efficient and accurate recommendation systems for e-commerce sites, streaming services, and content providers. By analyzing user behavior patterns, these systems can suggest products, movies, or music that users are likely to enjoy, enhancing user engagement and satisfaction.

- **Healthcare**: In healthcare, pattern mining can be used to identify common sequences of symptoms, diagnoses, and treatments that occur across patient groups. This can help to predict disease outbreaks, understand disease comorbidities, and improve the accuracy of diagnostic tools.

- **Fraud detection and security**: Banking and financial institutions leverage FPM to detect unusual patterns of behavior that may indicate fraudulent activities. This can involve identifying uncommon combinations of transactions, account activities, or claims that deviate from the norm.

- **Manufacturing**: In manufacturing, pattern mining can help identify frequent sequences of machine use or common patterns of system failures. This information is crucial for predictive maintenance, quality control, and optimizing production processes to reduce downtime and increase efficiency.

- **Web usage mining**: Pattern mining techniques are employed to analyze user navigation patterns on websites. This can help web designers optimize site architecture, enhance a user interface, and create a more intuitive user experience by understanding the most commonly accessed paths through the site.

- **Bioinformatics**: Pattern mining is used in bioinformatics for gene sequence analysis and protein structure prediction. It helps to identify common sequences or structures, which can be critical in understanding gene functions, evolutionary biology, and disease mechanisms.

- **Telecommunications**: In the telecommunications industry, FPM is used to understand common patterns in call records, usage data, and service faults. Insights gained can inform customer service improvements, network maintenance, and fraud detection in usage patterns.

- **Customer Relationship Management (CRM)**: CRM systems utilize pattern mining to analyze customer data and interaction histories to improve customer service, personalize marketing efforts, and enhance customer retention strategies by predicting customer behaviors and preferences.

- **Supply chain management**: Pattern mining helps to identify frequent patterns in supply chain data, such as common sequences of logistic activities or regular disruptions. This understanding can lead to more robust supply chain designs, better inventory management, and optimized logistics.

These applications demonstrate the versatility and value of FPM across different fields. By extracting and analyzing recurring patterns, organizations can gain actionable insights, foresee future trends, optimize operations, and deliver enhanced services or products to meet their users' needs.

FPM is a pivotal technique in data mining that has been applied across numerous domains, showcasing its versatility and effectiveness in extracting useful patterns from large datasets. Here are some of the most popular case studies demonstrating the application of FPM:

- **Walmart – market basket analysis**: One of the most classic examples is Walmart's use of market basket analysis to optimize its product placements and promotional strategies. By analyzing transaction data, Walmart can identify items frequently bought together and adjust their store layouts accordingly. This not only improves the customer shopping experience by making commonly bought items more accessible but also increases cross-selling opportunities.

- **Amazon – recommendation systems**: Amazon's recommendation engine uses item-to-item collaborative filtering, which can be considered an application of FPM. This system analyzes patterns of user interactions with products to recommend similar items. It has been a key driver in Amazon's ability to increase customer purchases through personalized recommendations.

- **Netflix – streaming content recommendations**: Netflix uses sophisticated algorithms that incorporate FPM to recommend movies and TV shows to users, based on their viewing history and the viewing patterns of similar users. This approach has significantly contributed to user engagement and satisfaction, making Netflix a leader in the streaming industry.

- **Target – predicting customer habits**: Target famously used FPM and predictive analytics to predict customer purchases, including sensitive cases such as predicting pregnancy, based on buying patterns. By identifying frequent patterns in purchase data, Target was able to send tailored coupons and advertisements to customers at the most opportune times.

- **Bank of America – fraud detection**: Financial institutions such as Bank of America use FPM to detect unusual patterns in transaction data that could indicate fraud. By identifying common patterns of legitimate transactions, a system can flag transactions that deviate significantly from these patterns, thus enhancing security measures.

- **Healthcare – disease outbreak prediction**: Public health organizations use pattern mining to analyze patterns in healthcare data, such as symptoms or geographic information, to predict disease outbreaks. For example, analyzing frequent patterns in patient visits and reported symptoms can help in the early identification and containment of outbreaks such as flu or COVID-19.

- **eBay – web usage mining**: eBay uses web usage mining to analyze the frequent navigation patterns of users on its website. Insights gained from these patterns help improve website design, enhance user experience, and optimize the placement of advertisements and promotions.

- **Google – search engine optimization**: Google uses pattern mining techniques to understand user behavior and preferences based on search queries and browsing histories. This information helps to refine search algorithms and enhances the relevance of search results and advertisements.

- **Spotify – music recommendations**: Spotify uses pattern mining to analyze user listening habits and the attributes of music tracks to recommend songs and artists that users might like. This not only keeps users engaged but also helps emerging artists get discovered more easily.

- **LinkedIn – professional connection recommendations**: LinkedIn utilizes pattern mining to recommend professional connections and job opportunities to users, based on their professional histories, interactions within the network, and patterns observed among similar users.

These case studies highlight the widespread applicability and effectiveness of FPM in deriving actionable insights, enhancing user experiences, optimizing operations, and driving strategic decision-making across various industries.

In the next section, we'll look at the challenges in FPM.

The key challenges in frequent pattern mining

FPM is a powerful tool for discovering relationships and structures in large datasets. However, implementing it effectively can be challenging, due to several inherent issues related to data complexity, scalability, and algorithmic efficiency. Here are some of the key challenges in FPM:

- **Scalability and efficiency**: One of the most significant challenges in FPM is managing the exponential growth of candidate itemsets as the size of data increases. The number of potential combinations of items grows exponentially with the number of items, making the mining process computationally expensive and time-consuming. Algorithms such as Apriori, which generate large candidate sets and check each for frequency, can become inefficient with large datasets.

- **High dimensionality**: In many real-world applications, datasets can have high dimensionality, meaning they contain a vast number of items or attributes. This high dimensionality exacerbates the problem of combinatorial explosion in candidate generation and can severely degrade the performance of the mining process.

- **Sparse data**: Many datasets in real applications, such as user-item interaction data in e-commerce or viewing histories in streaming services, are sparse. This sparsity means that many itemsets do not appear frequently, complicating the task of identifying meaningful patterns among a vast number of possible item combinations.

- **Noise and erroneous data**: Data quality issues such as noise and errors can significantly impact the outcome of FPM. Noise can lead to misleading associations and irrelevant patterns, which may be statistically significant due to their frequency but not meaningful in practice.

- **Handling the support dilemma**: Determining the right support threshold (that is, minimum frequency) for patterns to be considered "frequent" is a non-trivial task. A threshold that is too high might miss interesting patterns, while a threshold that is too low can produce an overwhelming number of patterns, many of which might be uninteresting or irrelevant. Balancing this to find the most insightful patterns without drowning in noise is a major challenge.

- **Incorporating domain knowledge**: In many cases, domain knowledge is crucial for interpreting the patterns discovered through mining. Incorporating this knowledge into the mining process to prioritize more relevant patterns or guide the exploration can be complex, but it is often necessary for actionable insights.

- **Pattern evaluation and interpretation**: Even after efficient mining of frequent itemsets, evaluating which patterns are truly interesting or useful remains a challenge. The subjective nature of "interestingness" and the context-specific interpretation of patterns means that additional criteria and post-processing are often needed.

- **Dynamic and evolving data**: In many modern applications, data is not static but continuously evolving. Adapting FPM algorithms to work with streaming data, where old patterns can become irrelevant and new patterns can emerge rapidly, poses additional challenges.

- **Privacy concerns**: Mining data to discover patterns can raise privacy concerns, especially when dealing with sensitive personal data. Ensuring that the mining process complies with privacy regulations and does not reveal individual user information is increasingly important and challenging.

- **Integration with decision support systems**: Lastly, integrating the results of FPM into decision support systems in a way that can be easily used and interpreted by decision-makers or end users is not straightforward. The usability of mined patterns in practical applications is crucial for their ultimate value to be realized.

Addressing these challenges requires sophisticated algorithm design, careful planning, and often a combination of multiple techniques and considerations to ensure that FPM is both feasible and effective in practical scenarios.

Frequent pattern mining algorithms

Several algorithms have been developed to efficiently discover frequent itemsets and association rules, even in large datasets. However, Apache Spark provides two algorithms, which we will discuss next.

FP-Growth

FP-Growth, short for **Frequent Pattern Growth**, is a robust data mining technique used to discover frequent patterns, associations, and relationships within large datasets. Unlike the Apriori algorithm, which is also used for FPM, FP-Growth requires only two scans of the transaction dataset, making it more efficient. It also does not need to generate candidate sets during frequent itemset generation, which saves computing resources and time.

The FP-Growth algorithm in Apache Spark has several valuable use cases. Let's explore the top three:

- **Market basket analysis**: FP-Growth is widely used for market basket analysis:

  - **Objective**: Discover associations between items frequently purchased together
  - **Example**: If a customer buys peanut butter, they are likely to buy jelly as well

  How FP-Growth helps:

  - It efficiently identifies frequent itemsets without generating candidate sets (unlike the Apriori algorithm)
  - It creates an FP-Tree data structure to represent item co-occurrences

- **Recommendation systems**: FP-Growth plays a crucial role in building recommendation engines:

  - **Objective**: Provide personalized recommendations to users
  - **Example**: Suggest related products based on a user's purchase history

  How FP-Growth helps:

  - It analyzes transaction data to find patterns
  - It generates association rules for item recommendations

- **Anomaly detection and fraud detection**: FP-Growth can be adapted for anomaly detection:

  - **Objective**: Identify unusual patterns or outliers
  - **Example**: Detect fraudulent credit card transactions

  How FP-Growth helps:

  - Finds infrequent itemsets that deviate from normal behavior
  - Unusual combinations of items may indicate anomalies

PrefixSpan

The PrefixSpan algorithm is designed to solve the problem of sequential pattern mining, which involves identifying frequent subsequences in a sequence database. Sequential pattern mining aims to discover all frequent subsequences (patterns) within a given set of sequences. A frequent subsequence is one that occurs with a frequency no less than a user-specified minimum support threshold. This task is challenging due to the potentially large number of possible subsequence patterns. Previous methods have typically followed an Apriori-like approach, which aims to reduce the number of combinations to examine but still faces challenges with large databases or numerous/long patterns.

The PrefixSpan algorithm in Apache Spark is a powerful tool for mining frequent sequential patterns. Let's explore its top use cases:

- **Sequential pattern mining**:

  - **Objective**: Discover frequent sequences of events or actions in a time-ordered dataset
  - **Example**: Analyzing clickstream data to find common navigation paths on a website

  How PrefixSpan helps:

  - PrefixSpan efficiently identifies sequential patterns by projecting prefixes of sequences
  - It avoids generating candidate sets explicitly, making it more scalable
  - It's useful for recommendation systems, process optimization, and behavior analysis

- **Session analysis and personalization**:

  - **Objective**: Understand user behavior during a session (for example, a website visit or app usage)
  - **Example**: Recommending relevant products based on a user's recent interactions

  How PrefixSpan helps:

  - It identifies frequent sequences of actions within a session
  - It enables personalized recommendations by capturing user preferences
  - It helps tailor content or offers based on recent behavior

- **Healthcare and medical research**:

  - **Objective**: Analyze patient data to find recurring medical patterns
  - **Example**: Identifying common symptom sequences or treatment pathways

  How PrefixSpan helps:

  - Mines frequent sequences from patient records
  - Aids in disease diagnosis, treatment planning, and epidemiological studies

Code examples on FPM

Let's review code examples using both the discussed algorithms.

Using the FP-Growth algorithm

Let's imagine a business use case that involves understanding which products are frequently purchased together by analyzing transactional data to determine associations between items. For instance, if customers often buy chips and salsa together, a retailer can strategically place these items near each other to encourage additional sales. Using frequent patterns, we can recommend related products (cross-selling) based on past purchases and create product bundles (for example, *"Buy this and get this free"* or *"Buy this and get a discount on another product"*) to increase sales.

The following code snippet imports all the required libraries and creates a Spark session:

```
from pyspark.sql.functions import monotonically_increasing_id
# Create a Spark session
spark = SparkSession.builder.appName(
    "NewsgroupsPreprocessing"
).getOrCreate()
spark.sparkContext.setLogLevel("ERROR")
```

The following code snippet loads the dataset and displays the DataFrame output:

```
store_data = spark.read.format('csv')\
        .option('header','true')\
        .option('inferSchema', 'true')\
        .option('timestamp', 'true')\
        .load('s3a://test234/store_data.csv')
store_data = store_data.withColumn(
    "id", monotonically_increasing_id() )
store_data.limit(5).toPandas()
```

Here's the output:

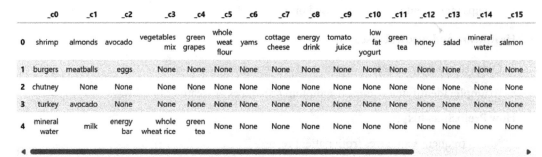

	_c0	_c1	_c2	_c3	_c4	_c5	_c6	_c7	_c8	_c9	_c10	_c11	_c12	_c13	_c14	_c15
0	shrimp	almonds	avocado	vegetables mix	green grapes	whole weat flour	yams	cottage cheese	energy drink	tomato juice	low fat yogurt	green tea	honey	salad	mineral water	salmon
1	burgers	meatballs	eggs	None	None	None	None	None	None	None	None	None	None	None	None	None
2	chutney	None	None	None	None	None	None	None	None	None	None	None	None	None	None	None
3	turkey	avocado	None	None	None	None	None	None	None	None	None	None	None	None	None	None
4	mineral water	milk	energy bar	whole wheat rice	green tea	None	None	None	None	None	None	None	None	None	None	None

Figure 8.2 – The dataset output through the DataFrame

The following code snippet preprocesses the data to align the data structure:

```
from pyspark.sql.functions import array,array_except,array_distinct
store_data = store_data.select(
    array(
        store_data._c0, store_data._c1, store_data._c2,
        store_data._c3, store_data._c4, store_data._c5,
        store_data._c6, store_data._c7, store_data._c8,
        store_data._c9, store_data._c10, store_data._c11,
        store_data._c12, store_data._c13, store_data._c14,
        store_data._c15, store_data._c16, store_data._c17,
        store_data._c18, store_data._c19
    ).alias("basket"))
store_data.show(5,truncate=True)
```

Let's see the output generated:

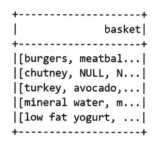

Figure 8.3 – The preprocessed data output

Next, the code sets the parameter for the FP-Growth algorithm. It then fits the data to generate the model. It displays frequent itemsets and association rules:

```
from pyspark.sql import SparkSession
from pyspark.ml.fpm import FPGrowth
# Initialize the FP-Growth model
fp_growth = FPGrowth(
    itemsCol="basket", minSupport=0.05, minConfidence=0.2)
# Fit the model to the transactions
model = fp_growth.fit(store_data)

# Get frequent itemsets
frequent_itemsets = model.freqItemsets
# Show the frequent itemsets
frequent_itemsets.show()
# Get association rules
association_rules = model.associationRules
```

```
# Show the association rules
association_rules.show()
```

The output of the frequent bought items is shown here:

```
+--------------------------+----+
|items                     |freq|
+--------------------------+----+
|[grated cheese]           |393 |
|[grated cheese, NULL]     |393 |
|[ground beef]             |737 |
|[ground beef, NULL]       |737 |
|[tomatoes]                |513 |
+--------------------------+----+
only showing top 5 rows
```

Figure 8.4 – A table showing frequently bought items

The following output shows the association rules by showing confidence, lift, and support for each item:

```
+--------------------+-----------+--------------------+--------------------+--------------------+
|          antecedent| consequent|          confidence|                lift|             support|
+--------------------+-----------+--------------------+--------------------+--------------------+
|              [milk]|     [null]|   1.0|1.0001333333333333|  0.12958272230369283|
|  [frozen vegetables]|    [null]|   1.0|1.0001333333333333|  0.09532062391681109|
|              [soup]|     [null]|   1.0|1.0001333333333333|  0.05052659645380616|
|  [whole wheat rice]|     [null]|   1.0|1.0001333333333333|0.058525529929342755|
|      [grated cheese]|    [null]|   1.0|1.0001333333333333|    0.0523930142647647|
|           [pancakes]|    [null]|   1.0|1.0001333333333333|  0.09505399280095987|
+--------------------+-----------+--------------------+--------------------+--------------------+
```

Figure 8.5 – A table showing the association rules

This completes our code walk-through of FPM using the FPGrowth algorithm.

Using the PrefixSpan algorithm

Let's imagine a business use case that involves identifying products frequently and sequentially bought together, by analyzing which products customers tend to purchase simultaneously and in sequence. This knowledge helps to tailor marketing efforts and improve recommendation systems.

Let's walk through the code example. The following code snippet imports all the required libraries and creates a Spark session:

```
from pyspark.sql import SparkSession
from pyspark.sql import Row
from pyspark.ml.fpm import PrefixSpan
# Create a Spark session
spark = SparkSession.builder.appName(
    "PrefixSpanExample"
).getOrCreate()
```

The next code snippet creates the synthetic data and displays the data through a DataFrame:

```
data = [
    Row(sequence=[["eggs", "bread"], ["apple"]]),
    Row(sequence=[["banana"], ["cheese", "chocolate"], ["eggs", "bread"]]),
    Row(sequence=[["eggs", "bread"], ["orange"]]),
    Row(sequence=[["milk"]])
]
df = spark.createDataFrame(data)
df.show(truncate=False)
```

We will see the following output:

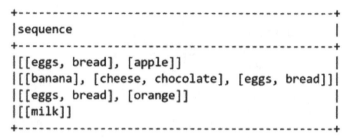

Figure 8.6 – The dataset output

The following code snippet sets the parameters for the PrefixSpan algorithm to find the frequent sequential patterns:

```
prefixSpan = PrefixSpan() \
            .setMinSupport(0.5) \
            .setMaxPatternLength(5) \
            .setMaxLocalProjDBSize(32000000)
result = prefixSpan.findFrequentSequentialPatterns(df) \
        .sort("sequence")
result.show(truncate=False)
```

Here's the output produced:

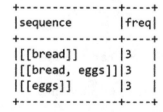

Figure 8.7 – The frequent sequential patterns output

With this, we have completed our walk-through of the FPM code example using the PrefixSpan algorithm.

Developing a model using scalable frequent pattern mining algorithms

Let's imagine the same business use case that we presented in the *Using the FP-Growth algorithm* section earlier in this chapter.

The following code snippet imports all the required libraries and creates the Spark session:

```
import numpy as np
import pandas as pd
import matplotlib.pyplot as plt
from pyspark.sql.functions import monotonically_increasing_id
spark = SparkSession.builder.appName(
    "NewsgroupsPreprocessing"
).getOrCreate()
spark.sparkContext.setLogLevel("ERROR")
```

The following code snippet loads and displays the data:

```
store_data = spark.read.format('csv')\
            .option('header','true')\
            .option('inferSchema', 'true')\
            .option('timestamp', 'true')\
            .load('s3a://test234/store_data.csv')
store_data = store_data.withColumn(
    "id", monotonically_increasing_id() )
store_data.limit(5).toPandas()
```

We then get the following output:

	shrimp	almonds	avocado	vegetables_mix	green_grapes	whole_weat_flour	yams	cottage_cheese	energy_drink	tomato_juice	...	green_tea	honey
0	burgers	meatballs	eggs	None	None	None	None	None	None	None	...	None	None
1	chutney	None	None	None	None	None	None	None	None	None	...	None	None
2	turkey	avocado	None	None	None	None	None	None	None	None	...	None	None
3	mineral water	milk	energy bar	whole wheat rice	green tea	None	None	None	None	None	...	None	None
4	low fat yogurt	None	None	None	None	None	None	None	None	None	...	None	None

5 rows × 21 columns

Figure 8.8 – The dataset overview

The following code snippet performs the data preprocessing:

```
from pyspark.sql.functions import array, array_except, array_distinct

store_data = store_data.select(
```

```
    array(
        store_data.shrimp, store_data.almonds,
        store_data.avocado,store_data.vegetables_mix,
        store_data.green_grapes,store_data.whole_weat_flour,
        store_data.yams,store_data.cottage_cheese,
        store_data.energy_drink,store_data.tomato_juice,
        store_data.low_fat_yogurt,store_data.green_tea,
        store_data.honey,store_data.salad,
        store_data.mineral_water,store_data.salmon,
        store_data.antioxydant_juice,store_data.frozen_smoothie,
        store_data.spinach,store_data.olive_oil
    ).alias("basket")
)
store_data.limit(5).show(truncate=False)
```

Let's see the output:

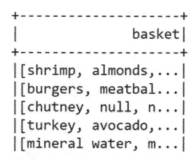

Figure 8.9 – The preprocessed data output

This concludes our walk-through of the code example.

Implementation in Apache Spark

In Apache Spark, metrics such as MSE and RMSE can be computed using the MLlib library, which includes support for ALS and evaluation utilities. Spark also supports creating custom evaluators if needed for specific metrics, such as precision@k or recall@k, which might require direct computation from the prediction results.

By utilizing these metrics, you can assess the performance of the ALS algorithm within Spark and refine your model to better suit the needs and preferences of your users, ultimately enhancing the effectiveness of your recommendation system.

Summary

In this chapter, we extensively explored FPM, a fundamental aspect of data mining. We started by understanding the basic concepts and importance of frequent patterns, highlighting their usefulness in discovering associations and correlations in large datasets across different domains. We delved into FPM algorithms such as FP-Growth and PrefixSpan, providing insights into their applications, strengths, and limitations through detailed case studies.

It is crucial to evaluate model performance, and we covered essential metrics and techniques to assess the quality of mined patterns. Furthermore, we discussed various strategies to enhance model performance, ensuring the effectiveness and efficiency of FPM processes.

By mastering these techniques, you now have the skills to uncover hidden regularities in data, apply them to real-world scenarios such as market basket analysis and web log analysis, and improve decision-making processes. The knowledge gained from this chapter equips you to leverage FPM for insightful data analysis and contribute to the advancement of data-driven applications.

In the next chapter, we will look at model deployment mechanisms.

Part 4:
Model Deployment

In this part, you will explore the crucial phase of deploying machine learning models into production environments. Deployment is a pivotal step in the machine learning life cycle, transforming theoretical models into practical tools that deliver real-time predictions and drive business value. This part will guide you through the various considerations, strategies, and best practices required to successfully deploy, monitor, and manage machine learning models in production.

This part will cover the importance of model deployment, the intricacies of building machine learning pipelines, and the practical aspects of model serialization, storage, and deployment strategies. You will also learn about the ongoing processes involved in monitoring, managing, and optimizing deployed models to ensure they remain effective over time.

This part contains the following chapter:

- *Chapter 9, Deploying a Model*

9

Deploying a Model

Model deployment is a crucial phase in the machine learning life cycle that bridges the gap between model development and delivering actual value through data-driven decision-making. Deployment is the process by which a machine learning model is integrated into an existing production environment to make real-time predictions based on new data. Understanding the importance of model deployment reveals its pivotal role in operationalizing data insights and achieving the practical benefits of machine learning.

This chapter focuses on several considerations and strategies for deploying a model in production. In this chapter, we're going to cover the following main topics:

- Importance of model deployment
- Exploring ML pipelines
- Model serialization and storage
- Model deployment strategies
- Monitoring and management
- Scalability and performance optimization

By the end of the chapter, you will have learned about the challenges involved in deploying a model and how to overcome them.

Technical requirements

The code files for this chapter are available on GitHub at `https://github.com/PacktPublishing/Apache-Spark-for-Machine-Learning/tree/main/Chapter09`.

Importance of model deployment

Model deployment is a transformative phase in the machine learning life cycle that converts theoretical models into practical tools that drive decision-making and business processes. It involves not just technical implementation but a strategic alignment of machine learning capabilities with business goals and processes. Ensuring successful deployment is therefore not just about technical proficiency but also about integrating analytical insights into the fabric of organizational operations.

Here are several key aspects that highlight the importance of model deployment:

- **Realizing business value**: The primary goal of deploying a machine learning model is to derive practical and actionable insights that can influence business decisions and strategies. Without deployment, even the most sophisticated models remain theoretical and do not contribute to business objectives. Deployment allows models to impact real-world processes and outcomes, enhancing efficiency, reducing costs, and potentially generating new revenue streams.

- **Scaling of solutions**: Deployment involves integrating the model into business processes, enabling automated and scalable solutions that can handle large volumes of data and transactions that would be impractical for manual processing. This scalability ensures that machine learning benefits are not confined to isolated tests or small-scale scenarios but are extended across wider operational frameworks.

- **Feedback loop**: Deployed models provide a continuous flow of information by interacting with real-world data and generating predictions or outcomes that can be measured and evaluated against actual results. This feedback is crucial for understanding the effectiveness of the model and provides data that can be used to further refine and optimize the model's performance. Continuous learning from incremental data feeds the model's adaptive capacity, enhancing its accuracy and relevance over time.

- **Competitive advantage**: In today's data-driven economy, the ability to quickly and effectively deploy machine learning models can provide a significant competitive edge. Deployed models can enhance customer experiences, streamline operations, and adapt to changing market dynamics faster than competitors who rely solely on traditional decision-making processes.

- **Regulatory compliance and risk management**: In many industries, machine learning models are deployed to ensure compliance with regulatory requirements or to manage risk. For example, in finance, models are used to detect fraudulent transactions in real time or to assess credit risk. Deployment in such contexts is not just beneficial but essential for meeting legal and operational standards.

- **Dynamic adaptation**: The deployment phase allows models to operate in dynamic environments where data patterns can shift unpredictably. By continuously monitoring model performance and retraining models with new data, businesses can adapt to changes more effectively. This dynamic adaptation is crucial for maintaining the relevance and accuracy of machine learning solutions.

- **Integration with technology stack**: Effective model deployment requires integration with the existing technology stack, which includes data pipelines, database systems, application servers, and possibly customer-facing applications. This integration is crucial for the seamless operation of the model within business processes and for ensuring that the model's insights are actionable and accessible to end-users or decision-makers.

Pre-deployment considerations

Selecting the right model for deployment involves a careful balance between technical performance, scalability, business needs, and operational constraints. A systematic evaluation based on these criteria helps ensure that the deployed model achieves the desired business outcomes while remaining cost-effective and maintainable over its operational life cycle.

Here are several key considerations to guide the selection of an appropriate model for deployment:

- **Business requirements**: Start by clearly understanding and defining the business goals:

  - **Objective alignment**: Ensure the model directly contributes to achieving specific business objectives, such as reducing costs, increasing revenue, improving customer satisfaction, or complying with regulatory requirements.

  - **Decision support**: Evaluate how the model's outputs will be used in decision-making processes. Consider the kind of predictions needed, the actions that will follow model predictions, and how these impact business strategies.

- **Model performance**: Assessing model performance involves more than looking at accuracy metrics:

  - **Accuracy and error metrics**: Choose the model that best meets the project's accuracy requirements while maintaining acceptable levels for other metrics such as precision, recall, F1-score, and so on, depending on the problem at hand.

  - **Validation techniques**: Use robust validation techniques such as cross-validation to ensure that the model generalizes well to unseen data.

  - **Performance trade-offs**: Sometimes, a simpler model with slightly lower accuracy might be preferable if it is significantly faster or more cost-effective in deployment.

- **Scalability**: Consider how well the model can be scaled to meet the needs of the business:

  - **Data volume and velocity**: Assess the model's ability to handle the volume and velocity of data expected in a production environment. This includes its performance in both batch processing and real-time predictions.

  - **Resource efficiency**: Evaluate the computational resources required by the model (CPU, memory, and storage) and whether these are feasible within the operational environment.

  - **Parallelization and distribution**: Consider models that leverage distributed computing frameworks such as Apache Spark if scalability is a critical factor.

- **Integration capability**: The ease with which a model can be integrated into existing business processes and IT infrastructure is crucial:

  - **Data integration**: Ensure the model can easily integrate with the data sources and data pipelines already in place.

  - **Operational integration**: Evaluate the ease of embedding the model into the current business operations and workflow without significant changes to existing systems.

- **Maintainability and monitoring**: Deployed models require ongoing maintenance and monitoring:

  - **Model drift**: Plan for how you will monitor and address model drift over time as the underlying data and relationships change.

  - **Update procedures**: Consider how easily the model can be updated or retrained. This includes the availability of new data, the complexity of retraining, and the deployment of updated models.

- **Regulatory and ethical considerations**: Ensure compliance with regulatory requirements and ethical considerations:

  - **Data privacy**: Ensure the model complies with the data privacy laws that are relevant to the jurisdictions in which it operates.

  - **Fairness and bias**: Evaluate the model for potential biases and ensure that it operates fairly across different groups of users.

- **Cost**: Consider the total cost of deploying and maintaining the model, including the following:

  - **Development costs**: This accounts for the time and resources spent on developing and tuning the model.

  - **Operational costs**: These are the ongoing costs related to infrastructure, monitoring, and personnel required to maintain the model in production.

After looking into several considerations for selecting the right model, let us shift the focus to ML pipelines.

Exploring ML pipelines

ML pipelines provide a powerful way to create and manage machine learning workflows in Spark. They offer a uniform set of high-level APIs built on top of DataFrames, making it easier to combine multiple transformations and algorithms into a single pipeline. Here are the key concepts:

- **DataFrame**: The ML API uses DataFrames from Spark SQL as ML datasets. DataFrames can hold various data types, including text, feature vectors, true labels, and predictions.

- **Transformer**: A Transformer is an algorithm that transforms one DataFrame into another by appending one or more columns. For example, a feature transformer might map a text column to feature vectors, creating a new DataFrame.

- **Estimator**: An Estimator is an algorithm that fits a DataFrame to produce a Transformer. For instance, a learning algorithm is an Estimator to produce a model by training on a DataFrame.

- **Pipeline**: A pipeline chains multiple Transformers and Estimators together to specify an ML workflow. You define the order of transformations and training steps in the pipeline.

- **Parameter**: All Transformers and Estimators share a common API to specify parameters. Parameters allow you to customize the behavior of each component.

  For example, suppose you want to preprocess text data, encode categorical features, and train a regression model. You can create a pipeline with `StringIndexer` (for categorical encoding), `VectorAssembler` (to combine features), and `LinearRegression` (as the estimator).

- **Model selection (hyperparameter tuning)**: Pipelines make it easy to perform hyperparameter tuning by specifying different parameter values and evaluating their impact on the model.

Remember that ML pipelines in Spark are inspired by `scikit-learn` and provide a convenient way to build, tune, and inspect practical ML workflows.

Code example of building an ML pipeline

The following code snippet demonstrates several feature engineering functions that transform the data before building a regression model using the linear regression algorithm. It then showcases how to build an ML pipeline incorporating all of these functions and algorithms.

The following code snippet imports all the required libraries and creates a Spark session:

```
from pyspark.sql import SparkSession
from pyspark.ml.feature import (
    StringIndexer, OneHotEncoder, VectorAssembler)
from pyspark.ml.regression import LinearRegression
from pyspark.ml import Pipeline
spark = SparkSession.builder.appName(
    "MLPipelineExample"
).getOrCreate()
```

The next code snippet loads the data (note that this dataset is only a sample for demonstration and not a real dataset). Then the data is preprocessed using several feature engineering functions to prepare the data to build the model using linear regression:

```
data = spark.read.csv(
    "path/to/your/dataset.csv", header=True, inferSchema=True)
```

```
dest_indexer = StringIndexer(
    inputCol="destination", outputCol="dest_index")
pipeline = Pipeline(
    stages=[dest_indexer, dest_encoder, vec_assembler, lr])

dest_encoder = OneHotEncoder(
    inputCol="dest_index", outputCol="dest_vec")
vec_assembler = VectorAssembler(
    inputCols=["feature1", "feature2", "dest_vec"],
    outputCol="features")
lr = LinearRegression(featuresCol="features", labelCol="target")
```

The next code snippet builds the ML pipeline for the model using the `Pipeline` function:

```
pipeline = Pipeline(
    stages=[dest_indexer, dest_encoder, vec_assembler, lr])
model = pipeline.fit(data)
transformed_data = model.transform(data)
```

> **Note**
>
> This code example demonstrates the setting up of an ML pipeline to stack several processing steps as one step as part of the ML pipeline.

In the next section, we will understand how to store the model.

Model serialization and storage

After learning about ML pipelines that produce ML models as an output, we will now learn about concepts of model serialization before saving the model and also about storing the model on storage media.

Model serialization

Model serialization in Apache Spark refers to the process of saving a trained machine learning model to disk so that it can be used for prediction later, without the need to retrain it from scratch. This is particularly useful in production environments where models are trained on large datasets and need to be deployed for making predictions. Spark provides built-in functions for both saving and loading models, which facilitate easy serialization and ensure that models can be used reliably in different environments or at different times.

Serialization techniques in Spark

Serialization in Spark allows models to be saved in a format that captures both the model's structure and its learned parameters. Here's how it works:

- **Persistence**: Models trained using Spark MLlib can be persisted (or serialized) to the storage layer. This is crucial for production systems where you need to maintain a model over time, despite possible system restarts or the need to share the model across different systems.

- **Format**: Spark typically saves models in a directory structure that contains all necessary information, including metadata about the algorithm and the ParamMap (parameters). The data is stored in Parquet format, a highly efficient columnar storage format.

Saving models

To save a model in Spark, you can use the `save` method that is part of the `model` class. Here's a general approach for saving a model using Python code:

```
from pyspark.ml.classification import LogisticRegressionModel
model_path = "/path/to/save/model"
model.save(model_path)
```

This will save the model to the specified path, which can be on a local disk or a distributed filesystem such as HDFS. The saved model includes all the necessary information to deploy the model for predictions without requiring access to the original code or data used to train the model.

Loading models

To load a saved model, use the `load` method associated with the `model` class. Here is how you can load a previously saved model using Python code:

```
from pyspark.ml.classification import LogisticRegressionModel
model_path = "/path/to/save/model"
loaded_model = LogisticRegressionModel.load(model_path)
```

This loaded model is now ready to make predictions in the same way as the original model was before it was saved.

Use cases for serialization

Serialization can be applied in the following use cases:

- **Production deployment**: After training a model on a development environment or on a subset of the full data, the model can be serialized and moved to a production environment where it can be deserialized and used for making predictions on new data

- **Model versioning**: By saving models to different directories or with different names, you can keep versions of models, which is useful for rollback, testing, or compliance purposes

- **Distributed computing**: In environments where training is done on one cluster, but prediction is needed on different nodes or clusters, serialization allows the model to easily be transferred and used across different parts of the system

Considerations

When implementing serialization, the following considerations need to be taken into account:

- **Compatibility**: Ensure that the Spark version used for saving the model is compatible with the Spark version where the model will be loaded. Changes in Spark's API or in the underlying model representation could lead to compatibility issues.

- **Security**: When saving models that may contain sensitive information, consider the security of the storage medium. Encryption and access controls should be employed as necessary to protect the serialized data.

Model storage

Apache Spark provides flexible storage options for serialized models, making it highly adaptable to different environments and storage requirements. When you serialize and save models in Spark, you can choose from several storage backends, each with its own set of features and use cases. Here's a look at some of the most commonly used storage options:

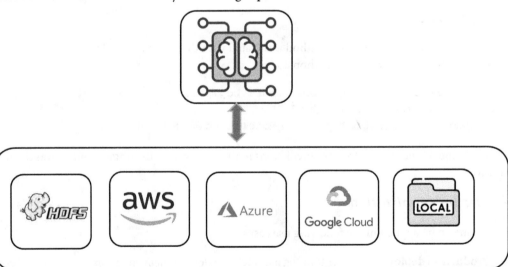

Figure 9.1 – Overview of model storage options

Let us look at several storage options available for storing the model:

- **Hadoop Distributed File System (HDFS)**: HDFS is a scalable and distributed filesystem designed to run on commodity hardware. It is particularly well-suited for use with Spark for several reasons:

  - **Fault tolerance**: HDFS provides high fault tolerance and is designed to be deployed on low-cost hardware. It stores multiple copies of data to ensure reliability and availability.

  - **Scalability**: It can scale up to thousands of nodes and petabytes of data, making it ideal for storing large machine learning models and datasets.

  - **Integration**: HDFS is natively integrated with the Hadoop ecosystem, which Spark is a part of, allowing seamless data processing and storage without needing to move or transform data extensively.

- **Amazon Simple Storage Service (S3)**: Amazon S3 is an object storage service offering industry-leading scalability, data availability, security, and performance. It's a popular choice for storing Spark models due to the following factors:

  - **Durability and availability**: Amazon S3 provides comprehensive security and compliance capabilities. Also, it stores data across multiple systems, ensuring high durability and availability.

  - **Cost-effectiveness**: With various storage classes and a pay-as-you-go model, S3 can be cost-effective, especially for data that doesn't need to be accessed frequently.

  - **Global accessibility**: Models stored in S3 can be accessed from anywhere, making it ideal for applications that need to serve users in different geographical locations.

- **Azure Blob Storage**: Azure Blob Storage is Microsoft's object storage solution for the cloud. It is optimized for storing large amounts of unstructured data, such as text or binary data, which includes machine learning models:

  - **Scalability**: It is designed to handle large amounts of data with no practical limit, making it suitable for enterprise-level applications.

  - **Security**: Azure provides strong security features that help protect data at rest and in transit.

  - **Integration**: It easily integrates with other Azure services, making it a preferred choice for businesses that are embedded in the Microsoft ecosystem.

- **Google Cloud Storage (GCS)**: GCS is a unified object storage for developers and enterprises, leveraging Google's infrastructure. It's a viable option for storing Spark models due to the following factors:

  - **High performance and scalability**: GCS is built on Google's core infrastructure and automatically scales to your business demands.

- **Data redundancy**: It offers various options for data redundancy and storage locations, which is crucial for disaster recovery and global accessibility.

- **Interoperability**: GCS supports the same unified API used by Google's other storage and database solutions, ensuring compatibility and ease of use across various services.

These advantages are in general applicable for all cloud service providers.

- **Local filesystem**: For development or smaller, non-distributed applications, Spark models can also be stored on a local filesystem. This is often simpler and quicker for development purposes due to the following factors:

  - **Ease of use**: Storing models locally can be managed without configuring network permissions or dealing with network latency.

  - **Quick access**: Access to local storage is usually faster than remote storage, which can speed up development and testing cycles.

Considerations

When choosing a storage option for Spark models, consider factors such as the following:

- **Data security**: Compliance with data governance and security policies

- **Access speed**: How quickly the data needs to be accessed

- **Cost**: Storage costs can vary widely based on the service and the access patterns

- **Geographical location**: Proximity to data centers can impact performance

Each storage option has its strengths and is suitable for different scenarios depending on the specific requirements of your Spark applications. When planning serialization and storage strategies, aligning these characteristics with your organizational needs is crucial for optimizing performance and cost.

Model deployment strategies

Model deployment strategy plays an important role in the success of ML projects. There are several ways of deploying the model based on the business requirement and context. Let us look at some of the popular options.

Batch scoring

Scheduling batch jobs for Apache Spark models to perform batch predictions on large datasets involves a series of steps to automate and manage the execution of model inference at specified intervals or in response to certain triggers. Here's a detailed guide on how to set up and schedule these batch jobs effectively:

1. **Define the batch job**: A batch job in the context of Apache Spark typically involves loading a trained model, preparing the data, making predictions, and then handling the output. This job can be encapsulated in a script that uses Spark's capabilities. You can write this script using Python (PySpark), Scala, or Java, which are supported by Spark.

2. **Prepare the Spark script**: Create a Spark script that performs the following tasks:

   - **Load the model**: Ensure that your model is accessible, either stored locally or on a distributed filesystem such as HDFS or cloud storage (AWS S3, Azure Blob Storage, and so on).

   - **Load the data**: Read the data on which predictions are to be made. This could be data stored in a database, a data lake, or live data being ingested at regular intervals.

   - **Data preprocessing**: Apply necessary transformations and feature engineering steps that the model expects as input.

   - **Run the model**: Use the model to predict the new data.

   - **Save/handle outputs**: Store the predictions in a database, send them to another service via API, or trigger another downstream process.

   Let's see an example of a PySpark script using Python code:

```python
from pyspark.sql import SparkSession
from pyspark.ml import PipelineModel
def batch_prediction_job():
    spark = SparkSession.builder \
        .appName("Batch Prediction Job").getOrCreate()
    model = PipelineModel.load("path/to/saved/model")
    data = spark.read.format(
        "parquet".load("path/to/input/data")
    predictions = model.transform(data)
    predictions.write.format("parquet") \
        .save("path/to/save/predictions")
    spark.stop()

if __name__ == "__main__":
    batch_prediction_job()
```

3. **Schedule the job**: To schedule and automate the execution of your Spark script, you can use various job scheduling tools. Some of the most common tools include the following:

- **Apache Airflow**: This is an open source tool designed to orchestrate complex computational workflows and data processing pipelines. Define a **Directed Acyclic Graph** (**DAG**) in Airflow to schedule and monitor your Spark jobs.

- **Cron jobs**: For simpler scheduling needs, you can set up a cron job on a Linux server where the Spark job is hosted. Cron jobs are useful for running scripts at regular intervals.

- **Apache Oozie**: A workflow scheduler for managing Hadoop jobs. Oozie can be used to schedule Spark jobs as part of larger workflows involving other Hadoop components.

Here's an example of the bash code of a cron job setup:

```
# Edit the crontab crontab -e # Add a line to run the script
every day at 3 AM 0 3 * * * /bin/spark-submit --master local[4]
/path/to/your_script.py
```

4. **Monitoring and logging**: Monitoring the execution of batch jobs is critical to ensure their correct functioning. Use logging within your Spark application to capture key events and outputs. Additionally, tools such as Airflow provide built-in monitoring capabilities that allow you to track job execution, retry failed jobs, and alert on failures.

5. **Optimization and maintenance**: Regularly review the performance of your batch jobs. Optimize resource allocation based on the job demands and update the workflows as your data or business requirements change.

By setting up scheduled batch jobs for Spark models, organizations can automate their prediction workflows, ensuring timely and efficient processing of large datasets for informed decision-making and operational efficiency. This setup not only saves time but also helps in leveraging big data analytics at scale.

Configure the scheduler

Let us look into various steps of configuring the scheduler in Apache Airflow:

1. Define a DAG that specifies the order of tasks and their dependencies.

2. Create a task to submit the Spark job, using `SparkSubmitOperator` or via `BashOperator` that invokes spark-submit.

3. Schedule the DAG by defining the execution frequency within the DAG definition.

Here's the corresponding code:

```
from airflow import DAG
from airflow.providers.apache.spark.operators.spark_submit import (
    SparkSubmitOperator)
```

```
from datetime import datetime
default_args = {
    'owner': 'airflow',
}
dag = DAG('spark_batch_prediction',
    default_args=default_args,
    schedule_interval='0 12 * * *',
    start_date=datetime(2021, 1, 1))

submit_job = SparkSubmitOperator(
    task_id='submit_prediction_job',
    application='/path/to/your-spark-job.jar',
    conn_id='spark_default',
    dag=dag)
```

RESTful API integration

Exposing a Spark ML model as a RESTful service allows it to be easily integrated into web applications and other systems that can consume REST services. This is particularly useful for deploying machine learning models to production, where they can be accessed and used for predictions from various client applications. Here's a step-by-step guide on how to expose a Spark ML model as a RESTful service:

1. **Prepare your Spark ML model**: First, ensure that your Spark ML model is trained, validated, and ready for deployment. You should have the model saved in a serialized format that can be loaded and used for predictions.

2. **Choose a web framework**: Select a web framework that you will use to create the RESTful API. Some popular Python web frameworks include Flask and FastAPI. These frameworks are lightweight and easy to use, and can work well with Python-based Spark applications.

3. **Set up the web server**: Set up a basic server using FastAPI, assuming your Spark environment is properly configured and your Spark session is active.

4. **Using FastAPI**: FastAPI is a modern, fast, and high-performance web framework for building APIs based on standard Python-type hints. It's especially good for asynchronous operations and when performance is critical:

```
from fastapi import FastAPI
from pydantic import BaseModel
from pyspark.ml import PipelineModel
from pyspark.sql import SparkSession
app = FastAPI()
spark = SparkSession.builder.appName("Model Serving") \
    .getOrCreate()
model = PipelineModel.load("/path/to/saved/model")
```

```
class PredictionRequest(BaseModel):
    features: list   # Adjust the request model based on your
feature set
@app.post("/predict")
async def predict(request: PredictionRequest):
    feature_df = spark.createDataFrame(
    [tuple(request.features)],
    ["feature1", "feature2", "feature3"])  # Adjust schema
accordingly
    prediction = model.transform(feature_df)
    prediction_result = prediction.collect()[0]["prediction"]
    return {"prediction": prediction_result}
```

5. **Containerize the application**: To deploy the application, it's a good practice to containerize it using Docker. This makes the deployment environment agnostic and simplifies scaling. Create a Dockerfile in the same directory as your FastAPI app:

```
# Use an official Python runtime as a parent image
FROM python:3.8-slim
# Set the working directory in the container
WORKDIR /usr/src/app

# Copy the current directory contents into the container at /
usr/src/app
COPY . .
# Install any needed packages specified in requirements.txt
RUN pip install --no-cache-dir -r requirements.txt
# Make port 5000 available to the world outside this container
EXPOSE 5000
# Define environment variable
ENV NAME World
# Run app.py when the container launches
CMD ["python", "app.py"]
```

Make sure you have a `requirements.txt` file that includes Flask and PySpark:

```
flask
```

Build and run your Docker container:

```
docker build -t spark-model-api .
docker run -p 4000:5000 spark-model-api
```

6. **Expose the endpoint**: Once your API is set up, expose the `/predict` endpoint to accept feature data and return predictions. Ensure that the API is properly secured and authenticated if required.

7. **Deploy the API**: Deploy your API on a server where it can be accessed by client applications. This could be a local server, a cloud-based instance (AWS EC2, Azure VMs, or Google Compute Engine), or a container orchestration service such as Kubernetes if scaling is needed.

8. **Testing and validation**: Test the RESTful API using tools such as Postman or write client-side code to make requests to your API. Validate the responses to ensure that the model is predicting correctly and the API is stable.

9. **Documentation**: Document your API endpoints, including the required request format and the structure of the response. Good documentation is crucial for end users to integrate with your API effectively.

Optional – monitoring and maintenance

Set up monitoring for your API to track its usage, performance, and health. Monitoring will help you identify and rectify issues such as downtime, slow response times, or errors in prediction.

By following these steps, you can successfully expose your Spark ML model as a RESTful service, making it accessible for real-time predictions in production environments. This setup allows your model to be consumed by various client applications, facilitating the broader integration and utilization of your machine learning solutions.

Automating model deployment pipeline

Deploying machine learning models in production requires careful consideration. Factors such as model type, size, expected impact, agility requirements, and hosting costs play a crucial role. Battle-tested strategies mitigate risks and optimize deployment speed. Furthermore, leveraging containerization, automated pipelines, cloud services, scalable infrastructure, and performance monitoring enhances deployment efficiency.

Creating automated pipelines for model deployment using Apache Spark involves several steps that ensure a seamless transition from model training to deployment and maintenance. This setup typically leverages **Continuous Integration and Continuous Deployment** (**CI/CD**) tools along with Apache Spark's capabilities to automate the training, evaluation, deployment, and updating of machine learning models.

Here's a step-by-step guide to setting up an automated pipeline for deploying Spark models:

1. **Environment setup**: Before you start, ensure that your development and production environments are properly set up with Apache Spark and any other dependencies needed for your project. This includes the following:

   - **Apache Spark installation**: Ensure Spark is installed and configured both in your development and production environments

- **Version control**: Set up a version control system, such as Git, to manage your codebase
- **Project structure**: Organize your project structure for Spark applications (include your training scripts, data schemas, and so on)

2. **Development**: Develop your Spark application locally. This includes the following:

- **Data preparation**: Scripts to preprocess and clean your data
- **Model training**: Spark MLlib scripts for training your model
- **Model evaluation**: Scripts to evaluate the performance of your model
- **Model serialization**: Code to save your trained model to a storage system
- **Versioning**: Implement model versioning to keep track of different versions of models trained at different times or with different sets of hyperparameters

3. **Deployment**: Deploy the model into a production environment. This can involve the following:

- Setting up a RESTful API layer as discussed earlier using Flask or another web framework
- Containerizing the deployment using Docker
- Deploying the container to a cloud service or an on-premise server

4. **Continuous integration setup**: Set up a CI system to automate the testing and building of your Spark applications. Tools such as Jenkins, CircleCI, or GitHub Actions can be used.

5. **CI server configuration**: Configure your CI server to monitor your version control system for changes.

6. **Build pipeline**: Create a build pipeline in your CI system that does the following:

- Checks out the code
- Runs unit tests and integration tests
- Packages the application (if necessary)

7. **Automated testing**: Implement automated tests that will run every time changes are pushed to your repository:

- **Unit tests**: Write tests for individual components of your application
- **Integration tests**: Test the integration between your application's components
- **Performance tests**: (Optionally) include tests to check the performance implications of changes

8. **Continuous deployment setup**: Set up a CD system to automate the deployment of your Spark application. This could be integrated with the same tools used for CI.

9. **Deployment scripts**: Write scripts to deploy the Spark application to a production environment. This often includes the following:

   - Pulling the latest model artifacts from storage

   - Deploying Spark jobs to a cluster

10. **Environment configuration**: Configure your production Spark environment to receive deployments.

11. **Automation**: Automate the deployment process to trigger on successful builds/tests from the CI pipeline.

12. **Monitoring and logging**: Implement monitoring and logging to ensure you can track the performance of your application and diagnose issues in production:

    - **Logging**: Ensure your application logs important events and errors

    - **Monitoring tools**: Use tools such as Prometheus, Grafana, or Spark's own monitoring tools to monitor the health and performance of your Spark jobs

13. **Feedback loop**: Set up mechanisms to feed information back from production to development.

14. **Model performance monitoring**: Regularly monitor model performance and accuracy.

15. **Data drift detection**: Implement tools to detect drift in your input data.

16. **Retraining pipeline**: Automate the retraining of your model if performance drops or significant data drift is detected.

    Here are some examples of how to set up the automated ML pipeline to deploy Spark models using Jenkins and Docker:

    - **Jenkins pipeline**: Define a Jenkins pipeline that builds your Spark job, runs tests, and deploys to production upon successful completion of tests.

    - **Docker**: Use Docker to containerize your Spark environment. This ensures consistency between different environments and simplifies deployment.

17. **Docker Compose**: Optionally use Docker Compose to manage your application stack, including Spark and any other services it interacts with.

As we saw, building automated pipelines for deploying Spark models involves integrating various tools and processes to ensure efficient, reliable, and maintainable model life cycle management. This pipeline not only automates routine tasks but also helps maintain the quality and reliability of your machine learning applications as they scale.

Model monitoring and management

There are several ways to monitor the performance of the model and management techniques to keep the performance in the acceptable range. Let us look at these techniques in detail.

Model performance monitoring

Monitoring the performance of machine learning models in production is crucial to ensure they continue to perform as expected over time. This process involves tracking key metrics, setting up dashboards for real-time monitoring, and configuring alerts to notify stakeholders of potential issues. Here's how you can implement these techniques effectively:

1. **Identify key performance metrics**: Before you can monitor anything, you need to define what metrics are important for your model. These could include the following:

   - **Accuracy, Precision, Recall, F1 Score**: Common for classification models
   - **Mean Absolute Error (MAE), Mean Squared Error (MSE), and Root Mean Squared Error (RMSE)**: Common for regression models
   - **AUC-ROC**: Useful for binary classification problems
   - **Business-specific metrics**: Examples include conversion rate, user engagement, or revenue impact, which reflect the actual impact of the model on business goals

2. **Data logging**: Implement logging mechanisms to capture predictions, actual outcomes, and input features. This data is essential for analyzing the model's behavior over time and diagnosing issues. Some of the data that could be logged include the following:

   - **Input features**: Log input features to later analyze which features are most predictive and to debug issues
   - **Predictions and outcomes**: Log both the predictions made by the model and the actual outcomes as they become known
   - **Timestamps**: Log when predictions were made to track changes over time

3. **Set up dashboards**: Use visualization tools such as Grafana or Kibana, or proprietary tools such as AWS CloudWatch or Google Cloud Monitoring, to create dashboards. These dashboards can display real-time metrics about the model's performance and health. The dashboard can display the following components:

   - **Real-time monitoring**: Display live data on model performance, including trends over time and distributions of predictions.
   - **Historical analysis**: Allow users to query historical data to identify patterns or changes in model performance.
   - **Dashboard metrics**: Include visualizations for all key metrics, such as accuracy or MSE over time, and histograms or boxplots of prediction scores.

Let's go through an example of Grafana setup:

```
# Assuming Prometheus as the data source
datasources:
  - name: Prometheus
    type: prometheus
    access: proxy
    url: http://your-prometheus-server
```

4. **Implement anomaly detection and alerts**: Set up anomaly detection to identify when the model's performance deviates significantly from expected patterns, which could indicate model drift, concept drift, or data quality issues. Some of the methods include the following:

  * **Threshold-based alerts**: Trigger alerts if key metrics fall below a certain threshold or change dramatically in a short time

  * **Statistical process control**: Use techniques such as control charts to monitor prediction performance and alert on unusual patterns based on statistical criteria

  * **ML-specific monitoring tools**: Consider tools such as Arize AI, Fiddler Labs, or WhyLabs, which specialize in monitoring ML models and can provide advanced analytics and automatic anomaly detection

Here's an example of Prometheus' alert rule:

```
groups:
- name: model-performance
  rules:
  - alert: HighPredictionLatency
    expr: predict_latency_seconds > 0.5
    for: 1m
    labels:
      severity: page
    annotations:
      summary: High prediction latency detected
      description: Prediction latency is above 0.5 seconds for
over 1 minute.
```

5. **Regular model evaluation**: Schedule regular intervals (for example, monthly or quarterly) to perform a thorough evaluation of the model against a recent dataset. Some of the options to evaluate the model are as follows:

  * **Retrain if necessary**: Depending on the evaluation results and how much the model's performance has degraded, decide whether to retrain the model with new data

  * **A/B testing**: Continuously test the current production model against newly trained models to determine whether a new model should replace the existing on.

6. **Feedback loop**: Establish mechanisms for capturing real-time feedback from users or downstream systems, which can be invaluable for assessing the model's performance and impact.

    - **User feedback**: Implement features in your application that allow users to provide feedback on predictions (for example, "Was this prediction helpful?")

    - **Feedback analysis**: Regularly analyze feedback to identify potential areas for improvement in the model

Model updating and maintenance

Maintaining and updating deployed machine learning models is crucial to ensure they continue to perform optimally as data evolves and business requirements change. Here are comprehensive strategies and best practices for updating and maintaining your models, including how to handle retraining and redeployment with minimal or no downtime:

- **Monitor model performance continuously**: Before considering updates, it's essential to have robust monitoring in place (as previously discussed). Continuous monitoring helps identify when a model's performance begins to degrade (model drift) or when external changes affect its accuracy. Key metrics to monitor include accuracy, precision, recall, RMSE, and business-specific KPIs. Alerts should be set up to notify the team when these metrics cross a predefined threshold.

- **Regular model evaluation**: Periodically evaluate the model using a current dataset to check its performance. These evaluations can be scheduled (for example, weekly or monthly) or triggered by significant changes in input data patterns or performance metrics.

- **Data management**: To effectively update models, the following is needed:

    - **Collect new data**: Continuously collect and label new data to refresh the training dataset. This data should reflect recent trends and changes in the environment that the model interacts with.

    - **Feature revision**: As new types of data become available or as the relevance of features changes, update the feature engineering steps to include or exclude features accordingly.

- **Retraining strategies**: The following are the two types of strategies for model retraining:

    - **Incremental training**: For models that support it (such as some decision trees, online learning algorithms), update the model incrementally with new data rather than retraining from scratch.

    - **Full retraining**: Periodically retrain the model with the complete updated dataset to incorporate new patterns and data fully. This approach is often necessary for complex models that don't support incremental updates.

- **Version control**: Use version control systems not only for code but also for models and their configurations. This practice helps in managing different versions of models and facilitates rollback to previous versions if needed.

- **Shadow deployment**: Before fully replacing an old model, be sure to do the following:

  - Deploy the new model in parallel with the existing one (shadow mode). This approach allows you to compare the new model's predictions with those of the currently deployed model under real-world conditions without affecting the system's output.

  - Use a traffic mirroring technique where the incoming requests are duplicated and sent to both the old and the new model, then log and compare the outcomes.

- **A/B testing**: Gradually expose a small percentage of users to the new model (A/B testing) and measure performance and user feedback against the currently deployed model. Increase the exposure as confidence in the new model grows.

- **Blue/green deployment**: Deploy the new version alongside the old version (green and blue environments). Once the new version is fully operational and tested, simply switch the traffic from old to new.

  This method reduces downtime and risk because if something fails, you can quickly revert to the old version.

- **Automate deployment process**: Use CI/CD pipelines for automatic testing and deployment. Automation ensures that the models are deployed in a controlled, repeatable manner, and reduces the human error factor.

- **Post-deployment monitoring**: Once a new model is deployed, monitor its performance closely. Ensure the monitoring system captures any unexpected behavior early. Adjustments might be necessary if the model performs differently in the production environment compared to the test environment.

In the next section, we will learn how to scale and optimize the model.

Scalability and performance optimization

It is very important to deploy a model that is both scalable and performance-optimized. Let us look at some of the ways in which we can achieve this.

Resource management

Effectively managing resources such as memory and CPU is critical when deploying models with Apache Spark, especially in production environments where efficiency and performance can directly impact operational costs and user satisfaction. Here are some best practices for managing Spark resources during model deployment.

Configure Spark properties wisely

Adjusting Spark configuration settings is crucial for optimizing resource usage:

- **Memory management**:

  - `spark.executor.memory`: This controls the amount of memory to use per executor process.

  - `spark.driver.memory`: This sets the amount of memory to use for the Spark driver process.

  It's essential to balance these settings based on the total memory available to avoid both out-of-memory errors and underutilization.

- **CPU utilization**:

  - `spark.executor.cores`: This specifies the number of cores to use on each executor. More cores per executor can lead to better throughput but might reduce parallelism if the number of executors is constrained.

  - `spark.default.parallelism`: Set this to an appropriate level based on your workload and the number of cores available. This setting determines the default number of partitions in RDDs returned by transformations such as `join` and `reduceByKey`.

Dynamic allocation

Enable dynamic allocation to allow Spark to dynamically adjust the number of executors based on the workload:

- `spark.dynamicAllocation.enabled`: Set to `true` to enable this feature

- `spark.dynamicAllocation.minExecutors` and `spark.dynamicAllocation.maxExecutors`: Configure the minimum and maximum number of executors

- `spark.dynamicAllocation.initialExecutors`: Provides a hint about the initial number of executors

Dynamic allocation is particularly useful for handling varying loads efficiently, as it allows Spark to free up resources when they are not needed and request resources when the load increases.

Persistent storage optimization

When your model deployment involves repeated access to the same data (for example, for batch predictions running at regular intervals), optimize data storage:

- Use data serialization to minimize memory usage. Configure `spark.serializer`. The Kryo serializer (`org.apache.spark.serializer.KryoSerializer`) is more efficient than the default Java serializer.

- Persist data using the appropriate storage level (`MEMORY_AND_DISK`, `MEMORY_ONLY`, `DISK_ONLY`, and so on) depending on your access patterns and memory constraints.

Garbage collection tuning

Tuning the garbage collector can help in managing memory more effectively and reduce pauses due to garbage collection.

Set `spark.executor.memoryOverhead` to account for non-heap memory usages such as Threads, JRE, and other native overheads.

Consider using different garbage collectors based on your application's requirements. For instance, G1GC might be more appropriate for applications with large heap sizes.

Resource allocation in the luster manager

If you're using a cluster manager (for example, YARN, Mesos, or Kubernetes), configure resource allocation settings to ensure that Spark applications do not monopolize cluster resources:

- **For YARN**: `spark.yarn.executor.memoryOverhead` specifies the amount of extra memory to be allocated per executor, above the value specified in `spark.executor.memory`

- **For Kubernetes**: Use resource requests and limits to specify the CPU and memory for each Spark executor pod

Fine-tune parallelism

Adjust the level of parallelism according to the data and the nature of the tasks.

Set `spark.sql.shuffle.partitions` according to your data size and operations. The default is 200, but you might need to increase or decrease this based on the data volume to optimize shuffle operations during tasks such as `groupBy`.

Monitor and optimize continuously

Utilize Spark's built-in web UIs to monitor the execution of Spark applications. Look for stages that take unusually long to complete or where data skew is causing imbalanced workloads across nodes. Iteratively refine your settings based on observed performance.

By adhering to these best practices, you can ensure that your Spark deployments are not only efficient in terms of resource usage but also maintain high performance, reliability, and cost-effectiveness in production environments.

Performance tuning

Optimizing the performance of deployed machine learning models in Apache Spark is essential for ensuring efficient data processing, especially when dealing with large datasets. Key techniques to enhance model performance include strategic data partitioning, effective use of caching, and fine-tuning other Spark configurations. Here's how you can implement these strategies.

Data partitioning

Partitioning in Spark controls the distribution of data across the cluster and has a significant impact on the performance of operations that involve shuffling data, such as joins, aggregations, and group operations. Some of the methods include the following:

- **Custom partitioning**: Use custom partitioners when you have knowledge about the distribution of your data that Spark doesn't. For example, if certain keys are known to be heavily used, partition data such that all records with the same key reside on the same node can minimize data shuffling during joins or aggregations.

  Here's a Python code example:

  ```
  from pyspark.sql.functions import col
  df = df.repartition(col("key_column"))
  ```

- **Managing partition size**: Adjust the number of partitions to avoid having too small or too large partitions. Small partitions can lead to excessive overhead in managing many small tasks, while too large partitions can cause memory issues and slow task performance.

  Sample Python code is as follows:

  ```
  # Repartition to a specific number of partitions
  df = df.repartition(100)
  # Coalesce partitions when reducing the number of partitions to
  minimize shuffling
  df = df.coalesce(50)
  ```

Caching and persistence

Caching data in memory can significantly improve the performance of your Spark jobs, especially if you need to access the same data multiple times for training or prediction.

- **Use cache wisely**: Decide which datasets need to be cached based on their reuse. Caching unnecessarily can lead to excessive memory usage and may evict other needed data from memory.

  Let's see an example Python code:

  ```
  df.cache()
  ```

- **Persistence levels**: Choose the appropriate storage level depending on your data access patterns and memory availability. For instance, use MEMORY_AND_DISK when you run out of memory so that Spark can spill data to disk instead of recomputing it.

  Here's the Python code:

  ```
  from pyspark.storagelevel import (
      StorageLevel df.persist(StorageLevel.MEMORY_AND_DISK))
  ```

Broadcast variables

For small reference datasets that are used frequently in Spark operations across multiple nodes, consider using broadcast variables. Broadcasting a dataset can optimize the performance of joins between a large dataset and a small dataset by keeping a copy of the small dataset on each node.

Explicitly *broadcast* smaller DataFrame to optimize join operations.

The Python code is as follows:

```python
from pyspark.sql.functions import broadcast
large_df.join(broadcast(small_df), "key")
```

Optimize serialization

Serialization plays a crucial role in performance, especially when data needs to be shuffled across the network or cached to disk.

Spark's **Kryo serialization** is faster and more compact than Java serialization. Enable Kryo and register your custom classes for the best results.

The Python code is as follows:

```python
spark.conf.set(
    "spark.serializer",
    "org.apache.spark.serializer.KryoSerializer")
spark.conf.set("spark.kryo.registrationRequired", "true")
```

Resource allocation

Properly configuring resource allocation is essential, especially in a shared environment.

Enable **dynamic resource allocation** to allow Spark to scale the number of executors based on workload.

Let's see the Python code:

```python
spark.conf.set("spark.dynamicAllocation.enabled", "true")
spark.conf.set("spark.dynamicAllocation.minExecutors", "1")
spark.conf.set("spark.dynamicAllocation.maxExecutors", "50")
```

Tuning Spark SQL

For deployments primarily using Spark SQL for data processing, use the following:

- **SQL configurations**: Tweak configurations such as `spark.sql.shuffle.partitions` to optimize the performance of operations that involve shuffling.

Here's the Python code:

```
spark.conf.set("spark.sql.shuffle.partitions", "200")
```

- **Cost-based optimizer**: Enable cost-based optimization for Spark SQL to allow Spark to pick the most efficient query execution plan based on data statistics.

    Here's the Python code:

```
spark.conf.set("spark.sql.cbo.enabled", "true")
```

By implementing these techniques, you can enhance the efficiency, speed, and scalability of your Spark machine learning deployments, ensuring that they handle large-scale data workloads effectively.

Summary

In this chapter, we delved into the comprehensive process of deploying machine learning models using Apache Spark, emphasizing its critical role in operationalizing data-driven insights and achieving practical business benefits. We began with pre-deployment considerations, highlighting the importance of model selection, data preparation, and training within the Spark ecosystem. We then explored model serialization and storage options, ensuring that models are preserved and accessible for future use.

We discussed various deployment strategies, including batch scoring, and the integration of models with RESTful APIs for broader application accessibility. The creation of automated deployment pipelines and best practices for model version control were outlined to streamline the deployment process and maintain consistency.

Furthermore, we addressed the essential aspects of monitoring and managing deployed models, focusing on performance monitoring, updating, and maintenance to ensure models remain accurate and effective over time. Scalability and performance optimization techniques were provided to manage resources efficiently and enhance model performance.

In conclusion, robust deployment practices are vital for the success of machine learning projects, enabling the transition from theoretical models to actionable tools that drive business value. By following the strategies and best practices outlined in this chapter, organizations can ensure that their machine learning models are effectively deployed, maintained, and optimized, leading to sustainable and impactful data-driven decision-making.

That brings us to the end. In this book, we explored the fascinating world of machine learning with Apache Spark. We delved into core concepts, leveraging Spark for powerful model development. Key takeaways include supervised and unsupervised learning, recommendation systems, and frequent pattern mining. Continue your journey, experimenting with algorithms and applying knowledge to real-world challenges.

Index

packtpub.com

Subscribe to our online digital library for full access to over 7,000 books and videos, as well as industry leading tools to help you plan your personal development and advance your career. For more information, please visit our website.

Why subscribe?

- Spend less time learning and more time coding with practical eBooks and Videos from over 4,000 industry professionals

- Improve your learning with Skill Plans built especially for you

- Get a free eBook or video every month

- Fully searchable for easy access to vital information

- Copy and paste, print, and bookmark content

Did you know that Packt offers eBook versions of every book published, with PDF and ePub files available? You can upgrade to the eBook version at packtpub.com and as a print book customer, you are entitled to a discount on the eBook copy. Get in touch with us at customercare@packtpub.com for more details.

At www.packtpub.com, you can also read a collection of free technical articles, sign up for a range of free newsletters, and receive exclusive discounts and offers on Packt books and eBooks.

Other Books You May Enjoy

If you enjoyed this book, you may be interested in these other books by Packt:

Mastering NLP from Foundations to LLMs

Lior Gazit, Meysam Ghaffari

ISBN: 978-1-80461-918-6

- Master the mathematical foundations of machine learning and NLP
- Implement advanced techniques for preprocessing text data and analysis Design ML-NLP systems in Python
- Model and classify text using traditional machine learning and deep learning methods
- Understand the theory and design of LLMs and their implementation for various applications in AI
- Explore NLP insights, trends, and expert opinions on its future direction and potential

Building Data-Driven Applications with LlamaIndex

Andrei Gheorghiu

ISBN: 978-1-83508-950-7

- Understand the LlamaIndex ecosystem and common use cases
- Master techniques to ingest and parse data from various sources into LlamaIndex
- Discover how to create optimized indexes tailored to your use cases
- Understand how to query LlamaIndex effectively and interpret responses
- Build an end-to-end interactive web application with LlamaIndex, Python, and Streamlit
- Customize a LlamaIndex configuration based on your project needs
- Predict costs and deal with potential privacy issues
- Deploy LlamaIndex applications that others can use

Packt is searching for authors like you

If you're interested in becoming an author for Packt, please visit `authors.packtpub.com` and apply today. We have worked with thousands of developers and tech professionals, just like you, to help them share their insight with the global tech community. You can make a general application, apply for a specific hot topic that we are recruiting an author for, or submit your own idea.

Share Your Thoughts

Now you've finished *Apache Spark for Machine Learning*, we'd love to hear your thoughts! Scan the QR code below to go straight to the Amazon review page for this book and share your feedback or leave a review on the site that you purchased it from.

https://packt.link/r/1-804-61816-0

Your review is important to us and the tech community and will help us make sure we're delivering excellent quality content.

Download a free PDF copy of this book

Thanks for purchasing this book!

Do you like to read on the go but are unable to carry your print books everywhere?

Is your eBook purchase not compatible with the device of your choice?

Don't worry, now with every Packt book you get a DRM-free PDF version of that book at no cost.

Read anywhere, any place, on any device. Search, copy, and paste code from your favorite technical books directly into your application.

The perks don't stop there, you can get exclusive access to discounts, newsletters, and great free content in your inbox daily

Follow these simple steps to get the benefits:

1. Scan the QR code or visit the link below

https://packt.link/free-ebook/9781804618165

2. Submit your proof of purchase
3. That's it! We'll send your free PDF and other benefits to your email directly

www.ingramcontent.com/pod-product-compliance
Lightning Source LLC
Chambersburg PA
CBHW080626060326
40690CB00021B/4834